The Perfect Stall

Karen E. N. Hayes, DVM, MS, PLLC
edited by René E. Riley

© Ironhorse Publishing
Hayden Lake, Idaho 2004

The Perfect Stall

Photographs by the author, with the exception of the photograph of the author on page 143, taken by Daniel Hayes.

Ironhorse Publishing LLC
21894 E. Hayden Lake Road
Hayden Lake, ID 83835 USA

Printed in USA

Library of Congress Cataloging-in-Publication Data
The Perfect Stall / by Karen E. N. Hayes, DVM, MS, PLLC — 1st edition.
Includes bibliographical references (p. 139-142).
ISBN 0-9747554-0-0
1. Horses. 2. Horses—Stabling. 3. Horses—Stalls. 4. Horses—Barns. 5. Horses—Care.
I. Title.

The Perfect Stall

Table of Contents

The Perfect Stall

In a perfect world, horses live in pasture.

In the real world, they live in stalls.

Make them Perfect Stalls.

Dedication

To Volkert,
the keeper of that chamber of my heart
that beats for horses.
My only regret is that he had to wait more than two
decades before finally getting his own Perfect Stall.

Once you've experienced

The Perfect Stall,

you'll never go back to the

old-fashioned way of

housing horses, no matter how

your granddaddy did it.

Introduction

The healthiest place for a horse is a clean pasture. In a perfect world, that's where all horses would be, all the time. But let's be practical. Most horses spend at least some time in a stall. When we stable our horses, we do the best we can to make their abode comfortable, safe, and healthy. But is it really the best we can do?

As an equine veterinarian I've seen every kind of horse facility, from the humblest backyard shack to the multimillion dollar show barn, from bare bones to mega-fancy. Believe it or not, the basic components in each stall are pretty much the same. In fact, we've been housing horses the same way for generations, for no other reason than it's the way it's always been done.

We horse lovers have a hard time stepping outside the box. In the process of making decisions from within that box, we're creating costly, career-busting, miserable, even life-threatening health problems for our horses. The dust, the ammonia fumes, the dark corners and shadowy light, the sneezing and coughing, the stinking mountain of waste behind the barn, the flies — these and other dirty little secrets go with the territory, right? It's just how horse facilities are. Right?

Not necessarily. The only reason we're still stabling horses the old-fashioned way is because as horsekeepers we've forgotten how to evolve. By our own doing, we're victims of tradition. *Tradition* is a polite way of saying, *stuck in the past*. The term *old-fashioned* puts a sentimental spin on the horse industry's refusal to grow and learn, even when it could save time and money to try a new approach.

I built The Perfect Stall because I finally got it through my own thick horseman's skull that it could be done—and within the average horseman's budget. I finally figured out I can give my horses a better place to live, and not only afford it but actually save money and time. I'm talking about *significant* savings.

For years I've thought about ways to improve things for my horses and my veterinary patients, and along the way I've come up with some effective but impractical methods that had little following. Well, I finally got it right. I now house my horses in stalls as fresh-smelling as my own living room. My farm no longer attracts flies, yet I use no insecticides or repellents. There's no "manure mountain" behind my barn, yet I don't pay anybody to haul the stuff away, nor do I have a complicated composting system with piles of urine-soaked bedding waiting for last month's pile to break down. My barn is fire-safe, and I plan to seek a reduction in my fire insurance premium (wish me luck). My stalls actually enhance my horses' health rather than put it at risk. I spend about one-fourth the time mucking stalls every day, and they have absolutely no ammonia smell and significantly less dust than they had before. And because

it's not necessary to tear down your old stall and start from scratch, you actually can have Perfect Stalls for your horses even if you're boarding in somebody else's barn.

In the chapters that follow, I'll describe old-fashioned stall components you're probably using right now. I'll show you why they're less than perfect (a lot less), and what they're doing to you, your horse, your schedule, your checkbook, and the environment.

Then I'll show you how I built my Perfect Stall, and how it can work to keep your horses healthier than they could ever be in an old-fashioned stall. And, I'll show you how keeping The Perfect Stall bedded, fresh smelling, and clean (in other words, how taking better care of your horse) will take a fraction of the time and money you spend, year after year, on your old-fashioned stall.

I'll show you that if you can afford to stall your horse the old-fashioned way, then you can afford to do it the Perfect way. In fact, you'll save money by installing Perfect components, whether you're building a stable from scratch, renovating an existing one, or renting a stall in somebody else's barn. More importantly, you'll provide your horse with a living space that'll promote his health and likely allow him a longer and more productive life. And, because you'll be removing the health risks that are prevalent in the old-fashioned stall, you'll likely be opening up new possibilities for your horse to perform up to his as-yet-unseen potential.

Even the Perfect Stall must improve

To protect The Perfect Stall from becoming mired in a tradition of its own, I'll tell you how even the Perfect components might improve—and what's new in the market—on The Perfect Stall's website, www.theperfectstall.com. Have an idea that'll make next year's Perfect Stall even better than this year's? Send it to the site! Note that the vendors whose products are featured on the site are all aware that their presence there isn't guaranteed, so there's a built-in incentive to stay current.

And, let this be a heads-up for all you inventors out there. Most of this world's brilliant inventions are simple ideas that came from the minds of people who weren't caught up in tradition, didn't have big corporate backing, and weren't afraid to step up to the plate and take a swing. Everybody has good ideas. Everybody. The innovators of the products featured in this book all said the same thing: Developing their idea and watching it take off was one of the best choices they ever made. If you've got an idea, join the elite group of forward-thinkers that are working to make horse care a dynamic, evolving, winning venture. Who knows, your idea might turn up in the next edition of The Perfect Stall!

Note: The product recommendations made in this book are NOT paid endorsements. They represent my opinion, based on my own personal and professional experience with the products tested at my facility. Whether these products work for you is for you to decide.

Chapter 1
Stall Construction

Horses naturally want to eat and sleep well away from their toilet area. While some stabled horses have a reputation for being "stall pigs," urinating and defecating willy-nilly instead of in a designated corner, I believe even the messiest horse will drop his manure in a discrete toilet area if his stall is large enough. If there's nowhere he can put his manure without it being within sniffing distance while he's eating, what's the point? If you have the option, build your stalls large enough so your horses can eat and sleep with a full body length between their hindquarters and the nearest manure pile. I'll bet the barn's worst "stall pig" is actually a fastidious tenant who simply couldn't operate under cramped conditions.

There's another payoff: Your stalls will be easier for you to clean, because the manure won't be trampled and fractured into the bedding. It'll take less time for you to find and remove the manure, and the stall bedding will be cleaner, because the manure balls will be intact.

Bottom line: Size does matter. The larger the stall, the easier it is for your horse to live in it according to the way he was designed.

What's a badly designed stall?

It's a stall that was designed by a human who may be very clever with hammer and nails but who doesn't know how to think like a horse. Many fancy show barns are good examples. Their fanciness serves to aggrandize the owners, but it does nothing for the horses. A Perfect Stall can be very humble. You can have a fancy Perfect Stall, but fanciness in and of itself does not make for perfection.

Don't be put off by the fact that I'm starting this chapter with talk about making your stall bigger. If that's not an option for you, that's okay. There are things you can do to your existing stall that'll greatly improve your horse's living situation. It won't make an absolutely Perfect Stall, but it'll be vastly better than it was before you picked up this book.

If your horse's stall was built badly (and many are), the very place you put him for the sake of comfort, health, and safety may be a place where he's uncomfortable, anxious, and prone to sickness and injury.

To design The Perfect Stall, I considered the following horsey characteristics.

1. As prey animals, horses tend to engage their number one defense whenever they feel the least bit threatened: *react first, think later*. I mean no disrespect. This strategy has worked well enough for the species to keep it galloping the face of the earth for millenia. Horses are action-oriented creatures of habit, driven by hair-trigger impulses and instincts. Any deviation from their normal routine can make them feel the need to make a forced choice on the spot. That choice usually is to assume the worst and flee the scene if at all possible, to get out of harm's way in a hurry, and then, from a safe distance, to gather their thoughts and survey the scene from a less pressured vantage point. Only then does the average horse engage his brain and decide whether that squeaky wheelbarrow or that squalling barn cat really is a threat.
2. Horses are social animals, designed to live in a horse community, with the ability to see, and interact with, the rest of the herd. They have a specific chain of command, and every single member has a place in that hierarchy. Because they're accustomed to

companionship and rely on other members of the herd to act as watchdogs and protectors, horses loathe isolation. When isolated, they're prone to developing "stall vices," such as weaving, pacing, stall-walking, digging, and cribbing.

3. Horses have a lot of skin surface, which is attractive to biting and sucking insect pests. They also have no fingernails with which to scratch itches. Out of necessity, they learn to scratch themselves by rocking their body parts over whatever surfaces present themselves, often to the detriment of that surface and of their skin.

4. Horses have a generally high-strung and inquisitive nature with a natural aversion to spending any length of time in an enclosed area where they can't see the horizon and hear the subtle sounds of stealth. When confined in a manner that blocks these natural tendencies, they're prone to expressing

Case #1: Hilal

Hilal lived in a hot, humid location with an abundance of insect pests. Her eyes were besieged, and the watery tears and crusty discharge at the inner corners of her eyes attracted even more flies. As a result, her eyelids were irritated and itchy. In an attempt to satisfy the itch, she found a protruding nailhead that had gradually backed out of a board by about a quarter of an inch. One or two rubs later and Hilal's face had a nasty laceration. I was called to stitch the wound.

pent-up nervous energy, again through the stall vices mentioned in (2) above, as well as stall kicking, wood chewing, and general mischief such as fiddling with latches.

With these qualities in mind, check the sidebars for cases I attended in the past 24 years as veterinarian. They illustrate how a stall can be built in ways that confound a horse's nature, resulting in stress and injury to the horse, and costly damage to the facilities.

The Perfect stall design

The perfect size for your stalls will depend on the size of your horses. My Perfect Stalls are 17' x 20', which comfortably accommodates my 17-hand Friesian horses. Larger is better unless your horse has a medical condition that requires

Case #2: Genevieve

Genevieve was sent to a boarding stable as part of her weaning. Though understandably a little agitated from having been abruptly separated from her herd of origin and trailered to a strange location, she seemed to settle into her 12'x 12' stall and was munching quietly on her dinner when the staff left for the evening. The following morning she was wandering the aisle, having jumped over the 4½-foot doorway. She was examined for injuries, but had only a couple of hairless spots on the upper fronts of her hind legs as a result of her escapade. She was lucky she didn't fare the same fate as Case #3.

figure 1.1
The Perfect Stall shedrow style building.

figure 1.2
The Perfect Stall basic construction.

close confinement for minimal movement. As a rule of thumb, measure the body length of your largest horse, from top of tail to muzzle. The smallest dimension of your stalls should be at least two times that measurement. That way, your horse's natural aversion to being confined is less challenged, he's more likely to choose a designated toilet area in the corner that's farthest from his feed, and his manure is likely to be at least a body's length away from his hindquarters while he's eating. Plus, if he wheels around in a hurry, he's less likely to collide with stall appointments. Such a collision not only would escalate his sense of claustrophobia but also could result in injury to him and damage to your stall.

Case #3: Shotzy

One night there was a violent thunderstorm. In the morning a show mare named Shotzy was found dead in a heap outside her stall. She had apparently tried to jump out during the storm but had caught one hind leg on the top of the 4½-foot doorway. For insurance purposes, I was called to perform a post-mortem examination. I found that as she fell, the caught hind leg was pulled so far out of position that its internal main artery, the femoral artery, tore. Shotzy had bled to death internally.

My Perfect Stalls are in a shedrow style pole building (figure 1.1, page 18) rather than built as stalls within another building. The shedrow style allows for an abundance of sunlight and fresh air, a plus in our mild Pacific Northwest climate. If we lived in a hotter region, the extra shade afforded by having the stall within a larger building might've been more desirable. In areas where the winters are more severe than they are here, the extra shelter of having the stalls inside a barn might be an important feature.

Case #4: Khebir

Khebir was a stallion. "His" mares grazed the pasture adjacent to his stall, but occasionally wandered out of his view. When they did, Khebir was prone to kicking his stall walls. One day he kicked the wall so hard that his foot broke through the wall made of 2" x 6" oak boards lining the stall, which was in a metal pole barn. Khebir's foot continued through the broken boards and out between the metal panels. When he yanked it back, the edges of the metal panels cut both sides of his fetlock area deeply, severing one major artery. It was a full year and several surgeries later before he walked without a limp, and he was unable to breed that year because the injured leg was too painful.

The poles are 6" x 6" treated timbers set in concrete at the corners and bisecting the long sides (figure 1.2, page 19). The base is a built-up pad of base rock and gravel, so the stall will be high and dry despite spring runoff. This type of sub-floor also discourages digging because it's less likely than dirt or clay to give the digger the satisfaction of easy excavation. There are horizontal treated timbers set into the base around the perimeter and screwed to the

figure 1.3
Tongue-and-groove pine in galvanized steel channels, topped with galvanized pipe.

figure 1.4
Remove a single screw to access the boards for replacement.

posts, to prevent the development of foot-grabbing holes between the wall and the floor.

To isolate the treated wood from the horses, there are 2"x 12" pine planks affixed to the interior of the posts and perimeter timbers, providing 6" of freeboard to corral the bedding once

Case #5: Patches

Patches lived in a metal building that had been used to store hay. She was a stall walker, a common stall vice in horses that have a difficult time adjusting to confinement. Around and around the perimeter of the stall she walked, sometimes clockwise, sometimes counterclockwise. The owners kept her stall thickly bedded with shavings. As a result, they didn't notice that Patches had pounded a deep depression into the floor all around the stall's edges. Outside, the dirt floor and shavings had begun to spill out from under the edge of the barn's metal walls. One day, while pacing around the stall, Patches' left hind foot slipped out under the wall. When she jerked it back in, she cut her foot so deeply on the metal edge that her entire heel bulb and part of the hoof itself were sheared away. I stitched up her wound under general anesthesia. Patches was a "kicky" horse, and her injury was painful, so she had to be sedated every week for her bandage change. By the time her foot had healed and she was able to walk soundly, the owners had spent enough money on veterinary care to build 2 Perfect Stalls from scratch.

the subfloor was installed. The back wall, and the wall that isn't shared by another stall, are 5½' high, just 2" shy of my horses' measurement at the withers. That way, they can see out and get fresh air, but even if they were scared out of their wits for some reason, they know they *couldn't* jump out. The partition shared by the adjacent stall is 7' high, to prevent nose-touching and fluid exchange by sneezing or coughing. The walls are 2" x 6" kiln-dried tongue-and-groove pine set in galvanized steel channels, topped with a galvanized pipe to provide a smooth, rounded surface and a finished look (figure 1.3, page 22). I chose pine over hardwood because of its resiliency in the event of a stall kicker—hardwood is beautiful, but somewhat more brittle. The tongue-and-groove construction

Case #6: Gaeli

Two weeks after Gaeli arrived at her new home, her groom asked one of the barn crew to bring her to the bathing area. The mare charged through the stall door, knocking down the crew member. When they finally caught Gaeli and got her into the crossties at the bathing area, an 8" gouge was discovered along her torso on the right side. Back at her stall, it was easy to see what had happened. The employee hadn't opened the sliding door all the way, and the door latch had protruded into the mare's path. In fact, a strip of her skin was still hanging from the tip of the latch. The employee claimed a back injury and went on workman's compensation. The mare's wound was stitched and healed without further incident.

is strong, resilient, has no nails to back out and rip skin, and it's easy to repair. If a board is broken, I can replace it by removing one screw from the upper steel channel and lifting the channel off (figure 1.4, page 23).

figure 1.5
Finger latch.

Stall fronts

For the stall fronts, I investigated builders of prefabricated modules. My aim was to get a stall front from a builder whose off-the-rack design already had attractive features, but who demonstrated a willingness to customize and innovate according to our shared ideas. With horse nature in mind, I looked for design flaws as well as genius.

figure 1.6
Adjustable hinge.

I gave the job to Dennis Marion of Innovative Equine Systems (IES) in Windsor, California. Marion, who hasn't ridden since age 9, doesn't consider himself to be a horseman, but knows a lot about horse nature. More, perhaps, than he realizes. His daughter rides their two Thoroughbreds daily, and Dad is observant, smart, quality-conscious, innovative, concerned about his daughter's safety—and that of the horses she loves—and is a good listener. He's also not shy about speaking up if he feels a buyer is asking for a feature that isn't in the best interests of the horse and handler. As a result, he manufactures some of the safest stalls on the market today.

Doors

The Perfect Stall front I chose features a hinged door rather than a slider. That way, no matter who's leading out a horse, the doorway is never made smaller by an incompletely opened door. If a horse hurries through while the door is partially closed, he'll simply bump it open on his way out. IES uses safety finger latches (figure 1.5, page 26) rather than a protruding latch that could snag clothing or horseflesh. "They are significantly more expensive for us to use," Marion told me, "but they're infinitely safer. If you're in a stall and have to get out in a hurry, you access the latch the same way from the inside as you do from the outside, without having to reach over the door awkwardly. It's one area where we really don't like to cut corners and compromise because so many mishaps occur at the door." The finger latches are also impossible for a horse to open, as the access port is smaller than any equine body part.

Hinges

The door hinge turns on a stout, blunt, stainless-steel pin welded to a cam, which allows for easy adjustment, for perfect fit. The door itself, as well as the wings of the stall front, have an 8" grill above, made of heavy gauge steel bars welded on 3" centers onto the channel steel for tongue-and-groove lumber below (figure 1.6, page 26). The spacing between the grill bars isn't random—the bars are closer together than in many of the other brands I checked. This prevents equine body parts from going between the bars and becoming trapped. The doors can

figure 1.7
Feed door to one side of main door.

be opened all the way, flat against the front, and held open with optional magnets. Thanks to tough materials and a well conceived design, the Perfect Stall front is so strong and solid that it doesn't need a connecting bar across the doorway at floor level, as most stall fronts do. "That bar they use, across the floor of the doorway," says Marion, "is a way of making up for basic weakness in the front's design. And, it's another hazard."

Marion cited a recent case in which a horse that was being led into a stall "clanged his hoof" on the floor bar and startled himself, bolted through the door and caused injury to the handler and to himself, resulting in a court judgment against the builder.

Instead of relying on a floor bar for strength and stability, IES installs their stall front door posts on 10" steel baseplates, which are anchored to the foundation. The result is a stall front that is stronger than any of the others I investigated. One of my 1,600-pound mares has proved its strength by charging into the front to repel a passing horse. She made one heck of a racket, but the stall front was unimpressed.

Feed doors

Marion customized my stall fronts with feed doors. That way, there's no sparring with an eager horse (which I worry about if someone who's relatively new to horses is helping with the feeding), and there's minimal spillage.

Social / Security

I chose IES's "Rubicon" design because it encourages horses to put their heads out, safely, so they can see each other and socialize. This helps satisfy their herd characteristics. If something startles them, the 5½-foot-high door, only 2" shy of their height at the withers, convinces them it'd never be a potential escape route. (That's the best way to keep your horse from hurting himself in a failed attempt to jump out—simply eliminate that option from his menu.) All the steel is hot-dipped-galvanized for long life in the face of exposure to the elements.

Lumber

The lumber itself is another IES innovation. It is 1"x 6" high-density polyethylene, similar to that used in constructing decks, but thicker, according to Marion's specifications. It's virtually indestructible. (It's also available in a variety of colors.) Unlike wood, horses are disinclined to nibble on it, and because it's all encased in steel channelling, they never gain access to a tempting edge. The "lumber" is non-porous, so it won't absorb urine, saliva, nasal discharges, or any other fluids; it's easy to clean and disinfect; and it never has to be painted or refinished.

I installed the poly-lumber without first contacting Marion for instructions (it looked so simple!), and as a result I made a common mistake: I fitted the boards too tightly in their steel channels. As a result, when the sun hits them, there's not enough room for them to expand without bowing. When the sun moves away, the bow resolves, but I'm going to fix the problem once and for all by sliding the boards out of their channels, cutting about ¼" off their length, and sliding them back in. No big deal. Marion concealed his exasperation well.

The only real downside to Marion's poly-lumber—and the only reason I didn't use it for the other walls of my Perfect Stall—is the cost. It's expensive. But someday, I'll replace the pine lumber with it. I love the idea of being able to power-spray and disinfect non-porous walls that'll never be touched by a paintbrush.

Support

Service and support are an issue with any product, and both before and after joining in with this project, the folks at Innovative Equine Systems have been quick to return phone calls, easy to talk to, open minded, and very helpful.

Cost and contact information:

At press time, the production price for the Rubicon design "off the rack" was an introductory $795. Customizing (size and/or features) adds $200. If you want powder coating instead of the hot-dipped galvanized I chose, add another $305. And, if you want to use the polyethylene lumber instead of tongue-and-groove wood, the Rubicon front's total package lumber is $325. As prices are likely to change over time, be sure to get an estimate.

Stall fronts from Innovative Equine Systems are available by calling Dennis Marion at IES: (800) 888-9921, or go online at www.equinesystems.com.

Why The Perfect Stall Loves IES Stall Fronts

- Designed for safe equine socializing.
- Doorway height discourages escape attempts.
- Solid construction, no floor bar across doorway.
- Bars are on three-inch centers for horse safety.
- Finger latches increase horse and human safety.
- Adjustable hinges allow perfect door fit.
- Customizing available.

Chapter 2
Bedding

As health factors go, bedding has the lead role in assembling The Perfect Stall. In this chapter, we'll take a fresh look at what we've all considered to be the standard, clean horse bed. First, I'll get specific about what's wrong with the most common types of stall beddings. Then, I'll tell you how I chose to bed The Perfect Stall, and how it's working out. Prepare to be surprised.

What's wrong with my bedding?

If you're the traditional type, you're probably using wood by-products (shavings, sawdust, wood chips), or forage by-products (hay, straw) to bed your horse's stall. If you're the unconventional type, you might be using an alternative or

A challenge for you

Want to convince me the bedding you use for your horse's stall is clean? Okay, I dare you: Bed the stall the usual way, so it looks clean and fluffy.
Now lie down in it and roll around.

Hmmm. Why are you hesitating?

Perhaps you're worried about those itchy bumps that sometimes occur after you've handled hay or straw, or shavings or sawdust. Those are from "forage mites," microscopic bugs (like chiggers) that naturally reside in organic bedding. Some bite; others burrow into your skin. And here's a nice tidbit: Some of the itch is due to feces they excrete under your skin.

The bugs don't particularly care what species they're bugging. A human, a horse, it's all the same to them; whoever shows up first.

You've also probably noticed that even processed, heat-dried bedding that comes in a plastic wrapper leaves you feeling gritty and dirty.
That's from dust, natural oils, and other contaminants.

Speaking of dust, if you've handled the bedding with your hands, try not to rub your eyes until after you've washed. And if you do take my dare and lie down in the bedding, it's best to hold your breath—that dust is full of allergens.

processed bedding material, such as peat, shredded paper, or manufactured wood-based pellets that swell when moistened.

If you're like me, you switch back and forth between different types of beddings, because none address all your stabling issues. They're dusty, the barn stinks, they make your horse snort and cough, they make you sneeze, they're labor intensive, they require a lot of valuable storage space (both for the fresh bedding and for the soiled muck-out), they're flammable, and they don't do a very good job of keeping your horse comfortable, clean, and dry. That's all bad enough, but it's actually worse than that.

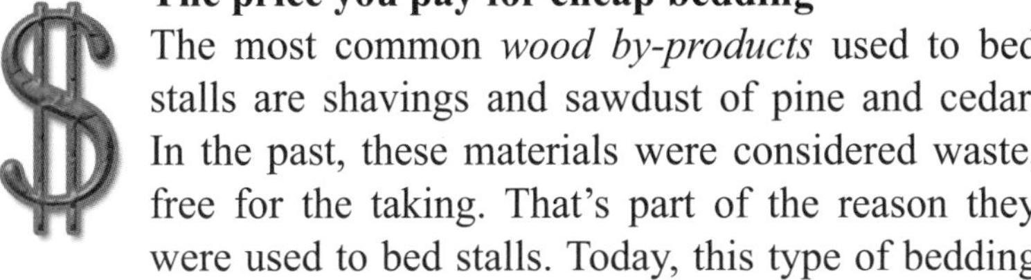

The price you pay for cheap bedding

The most common *wood by-products* used to bed stalls are shavings and sawdust of pine and cedar. In the past, these materials were considered waste, free for the taking. That's part of the reason they were used to bed stalls. Today, this type of bedding is far from free, because even when you don't have to pay for the bedding, you still have to pay somebody to deliver it to you, at a minimum of $60 per hour truck time. But, because horse and stable owners have gotten into the habit of using it, they continue to do so. And because it became habit to think of it as "free" or "cheap," the reality of what's recorded in the checkbook register never seems to sink in. Plus, pine and cedar smell nice. And, shavings and sawdust look fluffy and absorbent. But looks can be deceiving.

The other "cheap" beddings commonly used are *forages*—straw and hay. Selling straw gives the farmer a few extra bucks after he's harvested his grain. Hay sold as bedding (often because it's too old, dry, moldy, or weedy for feed) helps the hay farmer recoup some losses, and gives the horse a bed that can double as a midnight snack. Both straw and hay look fluffy, and you've been told they're absorbent.

So what's the problem? Well, there are 14 specific problems. A few of them probably won't surprise you. A few will.

1. Dust

No big surprise here. Kick a fresh pile of any of these beddings and you can see for yourself how dusty they are—even if they're "premium" quality.

You may be wondering, what's wrong with a little dust? Well, it's not a "little," and your horse spends a lot of time in his stall, breathing it in. The finer the dust, the deeper into his lungs and airways it goes. Dust may be organic, but that doesn't mean it's problem-free.

Dust is the number-one factor cited by researchers as a trigger for heaves, a career-killing lung disease that has no cure. Another name for heaves: *equine organic dust-induced asthma*. Hay and the bedding materials cited in this section are the most common source of the organic dust that has been associated with heaves, which can strike your horse at any time. Don't think he's immune just because he's reached a certain age. In fact, the older he is, the longer he's lived in that dusty environment and the greater the chance he'll be stricken. It's never too late to save your horse's lungs.

Even if your horse doesn't develop heaves, inhaled dust stimulates an inflammatory response in his lungs and airways, which react to the dust by producing pus and mucus to protect themselves. This gooey material partially clogs his "pipes," which narrow themselves even further by contracting in reactive spasms.

Your horse may or may not cough as a result. If he doesn't cough, don't be smug. Because of the inflammation and mucus, he's getting less air. As a result, even if you don't notice a difference, his

Airway Hyperresponsiveness

If your horse catches a respiratory bug, dust can delay his recovery. That's because for several weeks after he recovers from a respiratory infection, his airway is prone to a condition called *hyperresponsiveness*, where dust and other factors—cold air, smoke, chemical smells, and the heavy breathing associated with exercise—can throw him into asthmatic coughing fits. Coughing irritates the tissues even more. Airway hyperresponsiveness usually lasts about 6 weeks from the last time your horse coughed, so if you work him too soon, or put him in a dusty stall, or if there's smoke in the air from somebody's fireplace, and he has a coughing spell as a result, you've set the clock back another 6 weeks before he can return to work.

performance will be adversely affected. He won't run as fast, jump as high, last as long, or be as sharp as usual. In fact, you may never have seen his fullest potential, *so you may not even know what levels of performance he's capable of.*

Of even greater concern: As a result of all the dust-induced inflammation in your horse's respiratory tract, he's more vulnerable to illness than if he lived in a less dusty environment. He's less able to fight off *rhino* or *flu* or *strangles* or whatever other respiratory disease happens to be in the neighborhood, because his respiratory tract is chronically inflamed, which wears down his immune system.

The Cotton-Ball Test

Want to see how dusty your "clean" bedding is? After you've mucked out the dirty bedding, go into the bathroom and wash and dry your face and neck. Now go back to the barn and fill the stall in the usual way, with the usual fresh bedding. Fluff it up the way you usually do. Now go back to the bathroom mirror and soak a cotton ball with your favorite liquid facial cleanser. Wipe your face and neck thoroughly. Now take a look. That filth on the cotton ball is what stuck to your face and neck during the 10 minutes it took you to handle the so-called clean bedding. Imagine what your horse's skin, haircoat, nostrils, windpipe, eyeballs, and lungs look like after living in that stuff for even an hour, let alone overnight.... or 24/7.

What if you use an "alternative" bedding such as peat, or shredded paper?

Peat is pretty dusty stuff; the lower the quality, the dustier it is. It used to be a specialty item, relatively difficult to find, pretty expensive, and generally of higher quality than what's available today. Today you can buy it anywhere—at farm stores, home-improvement stores, and even grocery stores during gardening season, where it's stacked next to the plastic wading pools by the front door. Every vendor is trying to out-price the competition by buying their stock from ever-cheaper suppliers. As a result, virtually all peat available is the lower-quality, dusty stuff.

There's also an environmental issue with the mining of peat. Much of the peat on the North American continent is ancient, it's available only in finite supplies, and there are lots of vocal people who rightfully resent its removal from wetlands.

Shredded paper is dusty as well, and particles of paper dust tend to be very fine. The finer the dust, the deeper it penetrates into your horse's respiratory tract. Less dusty shredded paper usually comes from coated stock. The coating naturally makes it less absorbent; and, because paper isn't very absorbent to begin with, anything that makes it less absorbent is significant.

The bottom line: Among the most available and commonplace beddings, as well as the alternatives, dust is a big issue.

2. Mold

Organic beddings often contain actively growing mold and/or dormant mold spores. In fact, a portion of the dust discussed above actually is mold. If you don't pick up the typical "moldy" smell, that doesn't mean the mold isn't there—your nose isn't a very sensitive mold detector. Mold is highly allergenic (allergy-causing) for horses, just as it is for humans. And, it can worsen existing respiratory problems, such as heaves, asthma, and airway hyperresponsiveness (see page 39). Mold in hay bedding is doubly hazardous because it can cause serious colic if your horse eats it.

3. Wetness

Why do you put bedding in your horse's stall? Primarily, to keep your horse clean and dry. If your bedding is performing this duty well, then he should be able to lie down in his stall, sleep a while, roll around a bit, and be at least as clean and dry when he gets up as when he went down.

None of the beddings we've talked about thus far can make that claim, as anybody with a urine-stained horse will testify. That's because in a single 12-hour night, the average horse urinates two or three times, pouring a total of 2 to 6 gallons of urine into the bedding. Double that for the average 24-hour stay in the stall. Everybody knows it's unpleasant to be wet, especially to be wet with urine. And urine can be caustic stuff, hard on the skin.

It isn't the diaper that causes diaper rash.

Are some beddings more absorbent than others? Yes. Sawdust and manufactured wood-based pellets are probably the most absorbent of the bunch. Shavings: less good. Wood chips: even less. Forage bedding: worthless. Give peat high marks for being highly absorbent. Give shredded paper a failing grade.

The urine that got away

If you take 5 gallons of urine and spritz it in a fine mist over the entire surface of the stall's bedding, you might get even straw to absorb a good amount of it. But let's get real. Horses don't urinate in a mist, and it's not evenly distributed across the surface of the bedding—it's a sudden gush in two or three soggy locations. Even highly absorbent bedding has a tough time dealing with the sheer volume of it. As a result, the bedding gets soaked almost immediately, and much of the urine goes right through and makes puddles in low spots under the bedding, or seeps into cracks and crevices.

When you muck out the wet bedding, much of the "urine that got away" stays behind on (or soaked into) the stall floor. If you're innovative, you've probably learned the best way to remove urine puddles is to plop your horse's manure into them, smash it with your boot, then fork it out. *Your horse's own manure is more absorbent than the bedding.*

Hoof health

There's another problem with wet bedding, and this is an issue that a lot of people get wrong. *A dry environment is paramount for a healthy hoof.* In fact, the drier the environment, the better for the hoof. Standing in a moist bed weakens the structure of the hoof, just as a corrugated cardboard box weakens if it gets wet.

By the same token, excessive moisture also makes the hoof, and the cardboard box, less resilient. If you sit on the box when it's dry, it supports your weight, then springs back to

"...The drier the climate, the stronger is the horn of horses reared in it. The more upright are the hooves, and the more concave the soles. When the horn of the wall and sole is weak, it cannot sufficiently support the weight thrown on the leg, and the foot will have a tendency to become flat. The feet of horses bred in Australia, for instance, are stronger than those produced in England, owing to the climate there [in Australia] being drier. Under natural (dry) conditions, the hoof receives at least 99% of its moisture from the blood and lymph vessels. Yet it is more resilient, and often more moist, than that of horses kept in climates with periods of excessive moisture and periods of dryness. In fact, the most brittle, dry hooves are found in areas having wet winters and dry summers, such as western Washington and Oregon. The only domesticated horses whose feet generally retain a natural form and do not require hoof care are those kept and used heavily in arid climates."

--Captain M. Horace Hayes, veterinarian and horse expert; in: *Points of the Horse*, 7th ed. New York: Arco Publishing Co., 1968.

its original shape. If you kick it, it bounces away, unharmed. That's resilience.

In the same way, resilience, or elasticity, in your horse's hoof horn protects the internal structures of his hoof from injury due to concussion and heavy load-bearing. The externally dry, healthy hoof absorbs the abuse and springs back, thereby sparing his foot's internal structures.This isn't intuitive for most people. They think moisture equals elasticity. Well, they're partly right. However, the moisture that makes your horse's hooves elastic comes from *inside,* not *outside*. Here's what I mean.

It's an established fact that when a horse's hooves are chronically moist on the outside, the *periople* (the natural, varnish-like coating that protects the upper two-thirds of the hoof wall from over-evaporation of its *internal* moisture) gets damaged. This means, paradoxically, that when a horse's hooves are externally over-moist, they are more prone to becoming internally overdry. That's why applying external hoof dressings is discouraged, as is sanding and filing of the periople. When that same horse's hooves are chronically moist with *urine*, rather than plain water, it stands to reason that the damage to the periople will accelerate. Urine is so caustic that it can ruin leather and rubber boots. The result: weak, crumbly, misshapen hooves prone to fungal infection, that struggle to hold a horse shoe, and that tend toward underrun heels, flaring and breakage at the quarters, and flat, tender soles.

So, absorbency, or lack of same, is a test the old-fashioned beddings fail miserably. Even the most absorbent ones fail to keep your horse clean and dry. That's because after absorbing the urine, they hold it, like a soaked bath towel, willing and able to give it back when your horse stands in it or lies down.

4. Ammonia

Urine-soaked bedding is an ammonia factory. That's because urine contains the by-products of digested protein, which break down further with time and produce gaseous ammonia. Depending on how warm the ambient temperature is, it can take less than an hour for urine to start producing ammonia fumes—the warmer the stall, the faster the fumes are formed. Unless you muck out wet spots from your horse's stall every hour, there's plenty of time for ammonia emissions to build. In other words, the stall smells, and the smell gets worse with each passing hour.

Even after you muck out the wet bedding, the smell lingers, because the ammonia gas hovers in the air. If there's adequate ventilation (see Chapter 5), the gas should dissipate. But there's plenty more ammonia where that came from. Because most stall beddings are easily overwhelmed when your horse urinates, there's a considerable amount of urine that spills *through* the bedding and soaks into the stall floor, or seeps under the mats, or dribbles into the cracks between the floor and the wall. It accumulates there, cranking out a continuous supply of ammonia emissions. That's why the smell often is worse after you've mucked out the stall and allowed it to "air" for several hours. With every passing hour, more ammonia gas is produced from the "ammonia factory"—the urine that got away.

Ammonia fumes aren't just unpleasant, they're a health hazard. Breathing ammonia fumes can burn the eyes, nasal passages, throat, and lungs. The Occupational Safety and Health Administration (OSHA), dictates that workers exposed to 50 ppm (parts per million) ammonia gas mustn't be allowed to remain in that environment for longer than 2 hours in an 8-hour work day.

Let's put that into perspective. If you pour a tablespoon of nonsudsing household ammonia cleanser into a cup and measure the emissions at the mouth of the cup, they'll register at over 1,400 ppm. So, relatively speaking, 50 ppm isn't very much. However, your nose and eyes would disagree, as both are likely to be burning and watering when ammonia levels are at the 50 ppm mark. It's not uncommon for the ammonia levels in fancy, well appointed horse stables to be over 100 ppm. At 100 ppm, if OSHA had any jurisdiction over your horse's living conditions, he wouldn't be allowed to stay in his stall for longer than *half an hour* during each 24-hour period.

Where's the ammonia?

The strongest ammonia emissions in a horse barn are found where people generally aren't sniffing: in the stalls (not in the aisles), at a zone between the floor and your knees (not up by your nose when you're standing). Test it yourself. The next time you walk into your horse's barn on a warm day after all the stalls have been mucked out and re-bedded, take a whiff in the aisleway. Smell pretty fresh? OK, now open a stall door. Worse, eh? Now crouch down and take another whiff. Woo. Now put your nose right down about 3 to 6 inches above the bedding. Part the bedding a little and take another whiff. Every time your horse scuffs around in the stall, he uncovers a low-lying cloud of ammonia that lurks in the bedding's undertow. When his nostrils are down here, he's probably sleeping. That's when he's breathing the deepest, pulling the ammonia in.

A little perspective

Walk in the front door of a 2,500-square-foot house and odds are you can tell if they have a housecat. The ammonia fumes from the litterbox in the utility room get sucked up into the ventilation system and are wafted through the house. That's about a cup of daily urine output, maximum, in a 3-inch-deep pan of highly absorbent cat litter. Now think about the volume of urine a horse produces, in a 12' x 12' foot stall, and you can see why ammonia fumes are a significant issue for the horsekeeper.

With old fashioned beddings, there's nothing you can do about ammonia except muck out the soaked bedding, take your horse out and strip the stall, use some sort of stall freshener, and re-bed.

And, after all that, the ammonia smell returns and persists. There's always the urine that got away. Ammonia rules.

Even if you could eliminate ammonia fumes without removing the wet bedding, who'd want to do that? The bedding is soaked. You didn't want to lie down in it when it was dry, and I doubt if you're any more inclined to do so now that it's urine-soaked.

5. Hoof support and traction

Sawdust isn't bad in terms of hoof support. Peat is pretty good. Shavings and processed wood pellets are only so-so, and they lose a great deal of their loft and supporting ability when they get wet. Wood chips, shredded paper, hay, and straw: abominable, wet or dry. The larger the particles in any dimension, and the fluffier the medium, the less consistent contact the bedding makes with the sole and frog, and the less support they provide.

This lack of support,when combined with excessive moisture, can lead to loss of "cup" (concavity) of the sole—the sole becomes flat. This dulls the natural springiness of the hoof wall and limits the potential for expansion at the quarters. The flattened sole is similarly less elastic, which leaves a horse prone to smaller feet, and tends to make him tender-footed and uncomfortable when walking barefoot across uneven surfaces, such as gravel.

Bedding traction usually isn't a big issue while your horse is simply standing or walking around in the stall, but it can be a significant issue when he lies down, and even more so when he tries to get up. It's a major concern for the elderly, the arthritic, the obese, the weak, the lame, and the heavily pregnant. No matter the age or condition of the horse, in the final move to rise on slippery bedding it's not uncommon for his hind legs to slip out of position. The result: hock sores and sprains in the lower back and in the leg's major joints, including the hock, stifle, carpus, and fetlock.

6. Rest and comfort

Performance-horse trainers want their horses to lie down and rest, really rest. Deep sleep is essential to optimum performance, and the only way a horse gets deep, restorative, rapid-eye-movement (REM) sleep is when he's lying down. Some horses, in their entire adult lives, never get that. The so-called sleep they get while standing, with their locking-leg apparatus engaged, is a catnap at best, as they cannot be completely, totally relaxed—the locking apparatus will fail unless there's a slight amount of tension maintained. It's like sleeping in a passenger car on a long trip; you never wake up fully refreshed and ready to hit your stride.

Horses can survive without deep sleep. But thrive? That's another matter.

In pasture, where is your horse most inclined to lie down? Where he feels secure, where he's physically and psychologically comfortable, where it's dry, where he can hear the approach of potential danger, where he can get a good back scratching if he rolls, and where it'll be easy for him to get up in a hurry.

Horses are reluctant to lie down if doing so results in discomfort or a feeling of vulnerability. Straw or hay bedding provides so little support for bony protrusions that sleeping on it isn't much different than sleeping on the cold, hard ground. Unless bedded at least 8 inches deep, shavings are similarly deficient. And, the fluffier the bedding, the more a horse's heavy head sinks into it if he stretches out flat on his side, the most restful position. This makes it difficult for him to see, hear, and breathe without dust and bits of bedding wafting up his nose, triggering cough and sneeze reflexes that usually disturb the whole process, forcing him to his feet.

And, when fluffy bedding gets wet, it's not fluffy, lofty, or even marginally supportive any more. Lying down in formerly fluffy bedding after it gets wet is about as comfortable as lying down on a cold, wet tile floor.

Some horses are nervous enough, or low enough on the social totem pole, that they rarely feel comfortable enough *psychologically* to take the risk of lying down—it makes them feel too vulnerable. Bedding may or may not solve that problem, but it certainly can contribute to it.

7. Skin bumps

I've already mentioned the itchy little bumps that appear on your arms after carrying straw or hay, due to forage mites, which are similar to chiggers. Bugs can live in the pores of wood-waste bedding too, if it hasn't been heat-processed.

And, as you'll learn in the "Toxins!" section on page 55, wood-waste bedding contains natural, irritating oils and acids that can create a similar bump-ridden itch, even if there are no bugs.

Bedding-related skin problems in horses are particularly common on the cannon (shin) bones, fetlocks, and pasterns. And, bedding often is the cause of the occasional outbreak of "unexplained" hives in horses, either from direct contact with the skin, or from inhaling the bedding's dust. Among human workers in the paper, wood pulp, and sawmill industry, inhaling wood dust can lead to a skin reaction called atopy, from developing an allergy to some or all of the chemicals naturally present in the wood.

Imagine that I removed the smooth cotton sheets from your bed and replaced them with an itchy wool blanket beneath which I tossed a cup each of biting insects, grit, and tiny stickers and burrs from thistle-y plants. How well do you think you'd sleep, and in what condition do you think you'd find your skin in the morning?

8. Flies and the manure pile

Most people believe that flies are attracted to manure. The truth is, although many types of flies lay eggs in the manure, it's the smell of urine-soaked bedding that attracts them most strongly to the premises in the first place. When you muck your horse's stall and haul the manure and urine-soaked bedding out to the manure pile, you're creating a double-barreled neon welcome sign. It's sending out a tempting message to any fly within a mile's radius of the stable: an open invitation to come on over, and, once they've arrived, help themselves to a great big stinky pile in which to reproduce.

The manure pile itself is a target for zoning restrictions, homeowners' associations' rules, scrutiny and regulations by the state and local health department, and federal laws investigated and enforced by Confined Animal Feeding Operations (CAFO) regulations, and the Environmental Protection Agency (EPA). New horse facilities, especially those built near residential communities, watersheds, or environmentally sensitive areas, must submit a waste removal plan and adhere to it, or face legal action.

The average horse produces approximately 50 pounds of manure per day, and 75 pounds or more of urine-soaked bedding. This becomes a major disposal issue even for small, private facilities, where the pile becomes a monument to inefficiency in short order. It's a major financial concern for larger boarding stables and racetrack facilities, where hundreds of thousands of dollars are spent *per week* hauling manure and soiled bedding away.

The natural solution is composting, but depending on the facility, the muck pile can get too big, too fast, making it tough to maintain optimal composting conditions.

Meanwhile, the horses keep cranking out more, and more, and more manure and urine-soaked bedding.

Wood waste and forage bedding can ruin compost
In the case of wood waste, its natural chemical content, which is the basis for pine-oil cleansers and disinfectants, can kill the bacteria necessary for proper composting. Wood chips and shavings have been documented to resist decomposition. In fact, they resist it so well, they can remain intact in a compost pile for more than 10 years.

In the case of hay or straw, it's a common complaint that when mucking the stall, invariably too much bedding is removed. This not only grows the manure pile faster, it also pads the pile with tough, high-fiber material that takes a long time to break down, slowing the composting process. As a result, the bacteria necessary for a compost pile to work live out their normal lifespan—and die—before the job is done.

Your only remaining choices are to find a place on your premises that you can afford to set aside for waste disposal (and hope nobody reports you to the EPA), or haul the waste off the property (with associated trucking charges) to a facility that'll accept it (usually for a fee). If you can't certify that the manure is free of potentially harmful insecticides, medications, and herbicides, it could be considered hazardous material, which bumps the fee up. The bottom line: Much of the time, money, and real estate you spend on waste management could be better spent on enjoying your horses.

9. Fire

All the beddings mentioned in this chapter are a fire hazard, not only in the stall but also wherever you've stored fresh bedding. Even peat will smolder. Odds are, when you hear about a barn fire, you feel something that's about a hundred times worse than when you think about colic. That's a bad feeling.

As a veterinarian I've attended horses that survived a barn fire, and I've learned that burns are the least of their worries. It's the smoke. It does terrible things you can't see from the outside. Survivors might look pretty normal and lucky in the immediate aftermath of a fire, but check them the next day and you'll see they weren't as lucky as you thought. Smoke inhalation kills by cooking the horse's lungs.

In general, half the horses that die from a barn fire die during or immediately after the fire. The other half are put to sleep in the ensuing days. It's tragic, and it's preventable.

10. Storage

If, like many, you believe horse stall bedding needs to be light and fluffy in order to be effective and comfortable, then you know storing replacement bedding will take up a lot of space. At most horse facilities, space is at a premium, especially sheltered space.

While stored bedding sits around taking up space, it might also contribute to the overall dustiness of the premises, as well as to the risk of fire, especially if stored indoors. If you store it outside under a tarp, it's unsightly and growing mold fed by moisture that seeps down around the edges, and up from the ground. If it's wrapped in plastic bales, you can store it outside without worrying it'll get wet, unless there are punctures or rips in the plastic—any moisture that gets inside contributes to mold growth. And a stack of wrapped bedding bales is about as attractive as a stack of used tires.

11. Availability and cost

Wood-waste bedding is a declining industry due to shrinking availability and rising costs. Most mills now sign contracts to sell their chips, shavings, and sawdust to pressed wood, chipboard, and lignetics manufacturers. Even though there are still a few mills that give away their waste, you have to pay a trucker to bring it to your stable. And the cost never ends.

To bed a 12'x 12' stall 6 inches deep, and keep it fresh, can cost $1,200 per year per horse in bedding costs alone. That's not including the cost of labor. Even if you do the mucking yourself, it costs you valuable time to clean and re-bed that stall (which isn't clean even when the bedding is brand-new).

12. Labor

You haul fresh bedding in, and it takes several wheelbarrow loads to adequately bed the stall because the bedding is so *fluffy*—it takes up a lot of space. You haul it out wet, and it takes several loads because the soiled bedding is so *heavy*—2 gallons of urine alone weigh about 16 pounds. You dump the whole mess onto a pile. You haul in more dry bedding. The next day you haul out more soiled and soaked bedding. It's back-breaking work, time consuming, and it goes on forever.

The average muck load contains 80 percent urine-soaked bedding, and 20 percent manure, by weight. The average horse, in a single night in a stall, urinates two or three times (usually a gallon or two each time, at 8 pounds per gallon) and passes four to six manure piles weighing a total of about 50 pounds. This means that in the standard 12' x 12' stall, what you muck out after a single night's stay weighs 50 to 100 pounds, not including the bedding that comes out with the urine and manure.

.

13. Toxins!

In 2001, a sawmill in British Columbia was sued for killing fish in the Pacific Ocean. The mill lost the suit, paid a big fine, and had to clean up its act.

What had the sawmill done that was so toxic?

It left its pile of pine sawdust outside, where it got rained on.The rainwater trickled through the pile of sawdust and picked up toxic chemicals that are naturally present in pine, creating a dark "tea" known as **wood waste leachate**. The wood waste leachate spilled into the Pacific Ocean and killed enough fish to catch the attention of passersby. As it turns out, dead fish weren't the only problem. Those fish that got exposed but survived were rendered sterile. The impact on the environment was significant, and nobody knows how long it's been going on at other sawmills.

What did the sawmill have to do, to "clean up its act?" Keep its sawdust pile under a roof.

Who would've thought "clean" pine sawdust would be toxic to the environment? You're probably so familiar and comfortable with wood-waste bedding that you think of it as "natural" and "nontoxic." Well, it *is* natural. But it's far from nontoxic.

Not a walnut issue

You're probably already aware of the risk of black walnut in your horse's bedding. Simply standing in bedding that contains shreds of wood from a black walnut tree can founder your horse. Most suppliers of wood-waste bedding will *not* guarantee there's no black walnut in their product; if they do, they call it "horse shavings" and charge a premium price. If you're getting your shavings free from a cabinet maker or finish carpenter, rest assured this is *no bargain*, because hardwoods are virtually guaranteed to be in there. You might find black walnut, as well as exotic, oily woods of unknown safety from Brazil and other faraway places.

So let me be perfectly clear. When I say "toxic" wood waste, I'm talking about good old pine and cedar.

People who work in pine and cedar lumber, paper, and pulp mills are exposed to a lot of wood dust, perhaps even as much as a horse living in a stall with wood-waste bedding (and that's a lot). In one study, at least half those workers eventually were diagnosed with asthma. Those who had asthma before they started the job got significantly worse. At least half the people diagnosed with work-related asthma never recovered completely, even after they took up a different sort of job—years later, they still required at least intermittent treatment. The symptoms of asthma in people are remarkably similar to the symptoms of heaves in horses. Heaves is so common among stalled horses that researchers unanimously agree it's the stall environment that causes it.

We've already talked about dust (page 36), but it's more than that. Have you ever used a pine-based cleanser in your house and enjoyed that "clean" smell? Have you ever put cedar in your closet to keep moths out of your wool sweaters? At least some of the toxic elements responsible for the respiratory ailments in wood-based workers are also responsible for the natural aromatic and insecticidal properties of some cleansers and insecticides.

These chemical irritants are well-documented to cause cumulative damage to the delicate tissues of the respiratory tract over time. In cedar, the chemical is called plicatic acid. In pine, it's abeitic (a.k.a. sylvic) acid. In addition to asthma, workers complain of chronic coughing and have significantly reduced lung capacity. They also complain that they are thrown into fits of coughing and wheezing by many agents that didn't bother them before they were diagnosed with asthma—dust, smoke, chemical smells, perfumes, cold air, and the rapid breathing associated with exercise. Irritated nasal passages and irritated eyes are also common complaints.

Dust plus ammonia: a double whammy to lungs

By now you've probably noticed that the bedding itself can be as potentially harmful to your horse's respiratory tract, if not more so, as can the ammonia fumes from urine. Put the two hazards together and it should be no surprise that domestic horses have a lot of respiratory problems.

In fact, in a 2003 study done by the Pulmonary Laboratory at the Veterinary College at Michigan State University, 100% of seemingly healthy sport horses housed in a conventional stable environment were found to have significant inflammatory airway disease (IAD), and their age was not a factor—young horses (mean age 5 years) had just as much inflammation as did older horses (mean age 15 years). The source of the inflammation was hypothesized to be the dusty living conditions of the horses, who were valuable, well cared for, and high-performing athletes. The effects of their inflammatory respiratory conditions on their performance is likely to be significant, and yet we continue to house them in a manner that taxes their respiratory systems.

And, your horse may be more susceptible to infectious disease, due to chronically irritated airways. In fact, race track trainers vaccinate their horses every 2 months for "rhino" and "flu" because of the high rate of recurring respiratory infections. The horses' heightened susceptibility to infection, usually attributed to "stress," may be due at least in part to breathing ammonia, dust, and toxic chemicals naturally associated with wood-waste bedding.

Toxic things that go bump

Human studies have shown that the highly irritating and allergenic properties of wood-waste products can cause skin symptoms as well, resulting in some of those red, itchy bumps we talked about on page 49. In some cases it appears to be a contact allergy—the bumps develop on bare skin that was directly exposed to wood dust. In other cases, however, it appears to be a reaction known as *atopy*, where hives develop via an internal reaction after the dust is *inhaled.*

Unexplained hives are a common finding in horses bedded on wood-waste material. Hives are a serious and recurring problem in performance horse barns because the most common treatments—steroids and antihistamines—are forbidden at most events. The resultant dilemma drives competitors to distraction. Do you treat the hives and withdraw from the event? Or take your horse to the event, bumps and all?

And worse

In another human study there was a 300 percent increase in the risk of throat cancer among workers exposed to pine dust. Cancer of the esophagus is one of the more common epithelial cancers in horses. Is the incidence of this cancer in horses higher than it would be if they weren't living in stalls bedded with wood by-products? To my knowledge the research hasn't been done to answer that question.

14. Environmental and legal issues

The EPA has known for years that the toxic elements naturally present in wood products, including chemicals called resin acids, are environmentally hazardous. As a plank, apparently, wood poses little risk. It's when the wood is reduced to smaller particles that its toxic potential comes to bear. When water (such as rainwater or snowmelt) seeps slowly through a pile of sawdust or shavings, the resultant wood-waste leachate (see page 55) that percolates into the soil and contaminates the water table, or runs along the surface into the watershed, is rich in those toxic compounds. They're considered toxic to the environment, to any animal (including human) drinking affected water, and to plants that are irrigated with that water. They have been proven to sicken and kill fish, and interfere with their ability to reproduce. This is serious poison.

The United States Clean Water Act and EPA have beefed up the country's CAFO (see page 51) regulations to include moderate-sized horse stables on a case-by-case basis. It's like being audited by the IRS: A facility that is the subject of any sort of environmental complaint is likely to be put under a microscope regarding its environmental impact. Because of the way stables are typically managed, the result is not likely to be in favor of the horse facility.

The day is coming when each horse facility will have to get an operating permit from the EPA, requiring a written waste management plan and a pledge to store wood waste and equine waste in a covered location, in a manner that would prevent it from getting wet and thereby contaminating the environment.

The petitioner would also have to accept, in writing, legal and financial responsibility for environmental damages in the event of some natural disaster such as a 25-year storm washing toxins from the stable downstream or into the subterranean water table.

A good way to protect yourself against legal action down the road, if there were a severe storm that washed stable-waste contaminants into public watersheds, would be to have documentation that a government office, such as Soil Conservation and/or the EPA, signed off on your facility's plan, and to prove that management of the facility was done at least in accordance with, if not in excess of, published regulations.

If a sense of environmental responsibility doesn't cause you to re-think your love affair with shavings, the threat of a lawsuit just might. The way the CAFO laws are written, ordinary citizens are encouraged to bring suit against suspected offenders, on behalf of the local water table, public and private wells, lakes, streams, etc. As more of these suits appear in the archives, more litigious folks and organizations are eyeing horse facilities as the new target.

For generations, horse facilities have been unofficially excused, but if you plan to continue relying on that, then perhaps you haven't heard about the coffee roastery that was successfully sued recently for the pervasive smell of roasting coffee that wafted through the neighborhood.

Suffice to say it's no longer only the pig farmer who's considered offensive to the breathable air.

Speaking of noxious smells

Consider your manure pile. You've noticed flies and gnats buzzing around it. Here's how this contributes to environmental impact:

- The combined smells of the manure, the ammonia, and the rotting bedding can be considered a public nuisance.
- The fly population increases exponentially in response to the smell, which causes harried horsemen to engage in chemical warfare with insecticides and repellents.
- Pesticides kill the natural fly predators. The life cycle of the predators (the good bugs) takes longer than the life cycle of the flies (the bad bugs). The fly eggs hidden in the manure pile survive the fogging. They hatch and produce full grown, reproducing adult flies *faster* than do the predator eggs that survived the insecticide.
- Not to be outdone, some people feed their horses products that contain an insecticide, so their resulting manure is laced with fly-killing poison. What happens

to that poison when it rains, and the rain percolates through your manure pile? Does it soak into the ground, poisoning ground water? Does it run along the surface and into nearby streams, which spill into rivers and lakes and are absorbed by the plants and taken in by the fish and so on?

The Perfect Stall bedding

The bedding in The Perfect Stall is different—*very* different—from any other bedding available on the market today. At press time, I've used it for 2 years. Does it perform? Let's find out.

What It Is

It's called Equidry™ Bedding. It's such a departure from old-fashioned horse beddings that a lot of horsepeople find it difficult to consider. Their loss.

At first glance Equidry Bedding looks like cat box filler, but it's quite different. It's a red clay that's been kilned to over 2,000 degrees Fahrenheit for hardness, then milled into granules about the size of fish aquarium gravel. It's nearly indestructible—less than 1 percent of the product breaks down under a horse's weight per year. It was conceived by a geologist, who studied Anasazi pottery shards and marveled at how durable they were, persisting in the desert under tons of sand and emerging intact.

Equidry is a long-term bedding, not a disposable bedding. Now there's a concept that'll be tough for a lot of horsepeople to swallow. You install it, you remove the manure; the bedding stays put. What on earth will we do with all the time we usually spend mucking out urine-soaked bedding?

Absorbency

Each granule of Equidry Bedding is invested with hundreds of tiny nooks and crannies that vastly increase its surface area.

Figure 2.1
Stall bedded in Equidry.

Equidry and Sand Colic

In 2 years I've seen no evidence of any horses on Equidry making an attempt to eat it, and accidental ingestion should be minimal if the right kind of feeder is used (see chapter 3). If a horse did eat some significant amount of the bedding (and "significant" has yet to be defined in this instance), it's my belief that it'd pass through his system rather than accumulate the way sand could do. Why? Because Equidry is less dense than sand, and therefore less heavy. Sand settles out and irritates the gut lining. I believe the larger, lighter-weight particles of Equidry would be swept out with the manure. But I have no proof of that, and, as of this writing, the research hasn't been done.

Therefore, here's my recommendation: Most horses are disinclined to eat "foreign material," but once in a while I encounter an exception. I've attended a mare that ate a stud chain, for example. I've known more than one horse that ate baling twine. A tennis ball. A fist-sized rock from the driveway. A rubber curry. If your horse is so inclined, most bedding materials aren't for you, and that includes Equidry. Maybe your next horse will have a more discriminating palate and you'll get a chance to see just how great this stuff is.

That expanded surface area makes it amazingly absorbent. In one test, I dumped a 50-gallon barrel of water into a 12' by 12' concrete stall bedded in the standard 4 inches of Equidry. After an hour, there was *no puddle*.

In most cases, a minute or two after the bedding has been flooded (such as when two 5-gallon buckets of water are poured in one spot), the material grabs the liquid and holds it off the stall floor, but itself feels only damp, rather than soaking wet. That's because if the bedding is properly managed (I'll explain proper management on page 74), liquid is quickly wicked away from the point of origin to adjacent granules.

And, while other beddings absorb and *hold* moisture, which makes them soggy, no longer "soft and fluffy," and requires that they be removed from the stall and discarded, Equidry quickly re-evaporates the moisture it's absorbed, which it can do because of its peerless surface area. As a result, you remove only the manure. The urine evaporates, and you leave the Equidry in the stall.

Dust, mold, and other allergens

The product itself is dust-free, which is an enormous difference when compared to most other beddings. However, let's be realistic. A stall bedded in Equidry may not stay dust-free for long. That's because barns are typically dusty places, sporting dusty horses, dusty aisleways, dusty hay, dusty oats, and dusty tack and grooming tools. There also may be a dusty riding arena nearby, and dusty roads with dusty cars.

And, because the best barn is a well ventilated barn, there are open windows and doorways through which the wind blows more dust, and mold spores, and pollen. It all settles on everything, including the Equidry. Can anything be done about that?

Yes. Equidry Bedding can be *vacuumed* as effectively as can a shag carpet, which represents another great development for the barn environment as a whole. The company is spending as much (if not more) time developing their patented vacuum cleaner as they spent developing the Equidry itself. Here's how it works.

Think of your ancestors winnowing chaff from their grain by tossing it in the wind, and you've got an idea how this ingenious machine works. It agitates the bedding under a shrouded powerhead, which elevates dust and bits of lightweight fiber under the shroud, to be sucked up by the powerful vacuum, restoring the Equidry to nearly-new condition.

In my operation, vacuuming once every few months keeps the dust levels in the stalls (and in the overall barn environment) to an unprecedented low. I had one particularly abused stall, which hadn't been vacuumed for over a year and which had been contaminated with fragments of manure, hay, dirt, mud, dust, and all manner of filth from a construction project. (Construction workers were refurbishing my 100-year-old barn, and they were doing the sawing, cement mixing, and

The Straight-A Report Card

Absorbency: A+
My horses stay dry. Page 64.

Respiratory Health: A+
Equidry is inert, stable, non-allergenic and low-dust. Dust, mold, pollens, and any other fine inhalables that might have settled onto it from the environment are easily vacuumed. Page 66.

Time: A+
My horses stay in stalls 9 to 15 hours per 24-hour period. In the morning, all the muckout from six stalls fits into one wheelbarrow. It used to take three or four. Pages 66, 70.

Ammonia: A+
Now that I've figured out how best to maintain the bedding (there is a learning curve), there's virtually no ammonia smell. Page 72.

Hygiene & Flies: A+
Easy to clean. Easy to disinfect—Equidry is the only bedding that can be disinfected, with a patented, non-toxic disinfecting system. Fewer flies than ever on the premises, without insecticides or repellents. Pages 83, 84.

Safety: A+
Fireproof. Non-slippery. Pages 81, 84.

Comfort: A+
My horses love lying down in this bedding, for deep, more restful sleep. Arthritic horses love it. Page 79.

Hoof Health: A+
Quality of hoof horn has always been a problem with my imported horses, whose hooves were prone to crumble, crack, flare, split, and separate despite top-notch nutrition and farriery. On Equidry their hooves steadily improved, with healthier horn and restored cupping of sole. My farrier now has Equidry in his barn. Page 81.

Haircoat: A+
Even if my horses enter their stalls at night covered with mud, their smooth, glossy coats in the morning make it look as though I'm a compulsive groomer. Just between us horsemen, my horses rarely get groomed. They're cleaner than most humans. Page 82.

Cost: A+
You might gasp at the price tag at first, but remember, you're used to buying disposable bedding. This stuff is permanent. After 12 months, your bedding is paid for—and it'll be years before you'll need to top off the stalls. You'll also save time and money on labor. Page 70. And, you won't have to search for shavings when your supplier dries up.

Storage: A+
No pile of shavings to re-bed stalls, no straw. Page 84.

Environmental Issues: A+
Inert; nontoxic. No wood-waste leachate. No urine-soaked bedding pile. Reduced need for insecticides and repellents. Chemical-free disinfection system.

lunching in that particular stall.) A half hour with the vacuum system removed roughly 50 gallons of fibrous, smelly material, sawdust and fine dust, and undoubtedly a gallon or two of pulverized Cheetos® and strawberry Danishes, leaving the Equidry looking and smelling like new. Before vacuuming this stall, tossing a handful of the bedding into the air elicited an obvious cloud of dust. Afterwards: no visible dust at all.

Labor savings: do the math

Does vacuuming the stalls negate my daily time savings with Equidry? Not even close. It used to take 20 minutes or more to do my usual daily muckout of each stall, when the stalls were bedded in shavings. That included digging out wet spots and manure, hauling the manure and urine-soaked bedding to the dump pile, sprinkling stall freshener on the wet spots, hauling in a wheelbarrow of fresh shavings to replace what I removed, and spreading it around.

Now that I have Equidry, it takes less than 5 minutes to do the daily cleaning.

- 3 minutes or less to pick up the manure.
- 2 minutes or less to plug in and run an electric aerator through the stall.

In a 30-day month, that means I now spend a total of 2½ hours or less per month on muckout, per stall, instead of 10 hours per month, per stall. That's a time savings of 75 percent, or 7½ hours per stall per month.

Let's say that once a month, I vacuum the stalls. That's probably a bit excessive—in my part of the country, where the air is pretty moist and dust is less of an issue, I'm satisfied with vacuuming four times a year. But let's err on the side of caution. I spend a leisurely 20 minutes vacuuming each stall. If I had to do it every day I might squawk, but once a month it's a pleasure. Especially because the stalls, which were good-looking even before vacuuming, look absolutely fabulous afterwards.

That brings my monthly total (for daily muckout plus once-monthly vacuuming) up to a little less than 3 hours per stall per month. Compare that to the 10 hours per stall per month I spent before, when the stalls were bedded in shavings. (It'd take even longer if I used straw, which is a real pain to muck.)

That's a monthly time *savings* of 7 hours per stall.

I clean 8 stalls every day. Now that I have Equidry, I'm saving 56 hours per month, or 672 hours per year. At $6 an hour, Equidry saves $336 in wages, every month, or $4,032 a year. On a per-stall basis, that's a labor cost savings of $504 per stall per year. So, if I cleaned 10 stalls, my annual labor savings would be $5,040. (And I pay more than $6 per hour.) How much money do you make in an hour, on average? I suspect it's more than $6. If you clean your stalls yourself and add up the hours saved… well, you do the math.

Ammonia

In the laboratory, Equidry appears to be the perfect solution to the ammonia-fumes problem in horse stalls. In the real world, it's lived up to its promise. There's a learning curve, but I've already hiked that curve and am ready to share what I learned so you can have fresh, sweet smelling stalls every day, without exception.

The trick is to assess your facility's needs, then make sure you maintain your stalls in a manner that accommodates the local climate, the stalls' level of use, your horses' eating and toileting habits, etc. For more information on these issues and how to work with them, see pages 76, 77 and 79. First, a little background.

If they urinate, ammonia will come

It takes water, bacteria, and a little time to make ammonia gas from the protein by-products (principally urea) naturally present in urine. In a tray in the laboratory, Equidry was so good at evaporating water that by the time the urine began forming ammonia gas, its water was gone, which essentially put the ammonia factory out of business.

And, because each granule of Equidry contains hundreds of nooks and crannies filled with air, urine breakdown is more likely to take place *aerobically*— i.e., with oxygen, instead of *anaerobically* (without oxygen). Aerobic breakdown of organic substances is almost always associated with less stink, and that applies to ammonia as well as other smells.

Think of the wet garbage generated in your kitchen. If you separate it from your dry garbage and keep it in an airtight container, you've probably noticed that whenever you take the lid off to add a banana peel, the smell of rotting garbage is pretty strong. If, on the other hand, you leave the container uncovered, the smell is minimal.

Issues that'll affect how Equidry should be maintained

- Organic fines that get into the stall over time can absorb urine and remain wet, just as old-fashioned bedding does. This includes dust, hair, dirt from your horse's hooves, hay chaff, and spilled hay leaves and stems. When urine meets up with this organic chattel, it soaks in and stays there, hiding from the drying characteristics of the Equidry. The best way to prevent this is to pick out spilled hay with the manure fork daily, choose a Perfect Stall feeder that minimizes spillage (page 89), and keep the organic fines vacuumed up as often as necessary. How often? It depends on the size of the stall and on how much mess your horse makes each night. For my draft horses, I think once every few months is about right.

- Another issue that can lead to ammonia fumes is the *mulching effect*. This is a phenomenon that keeps a liquid layer from joining with an adjacent solid.

The Perfect Maintenance Routine

Here's my quick and easy muck routine to maintain a clean, fresh, lung-friendly, pleasant-smelling, suitable-for-visitors, Equidry bed in the Pacific Northwest.

You'll need the following tools:

- Equidry-brand manure fork/basket. The tines are closer together than on other brands, so you'll be more effective at catching smaller pieces of manure and spilled hay. And, because Equidry offers so little resistance to penetration by the fork, the tines plunge in easily.
- Wheelbarrow or muck bucket.
- Method of mixing the wet with the dry Equidry in the stall.

    I chose power agitation over doing it manually, using the electric aerator available through the Equidry company. It's easier and does a much better job than strong-arming the bedding.

Here's how it's done:

1. Remove the manure and spilled hay with the manure fork.

   Time: 1.5 to 3 minutes.

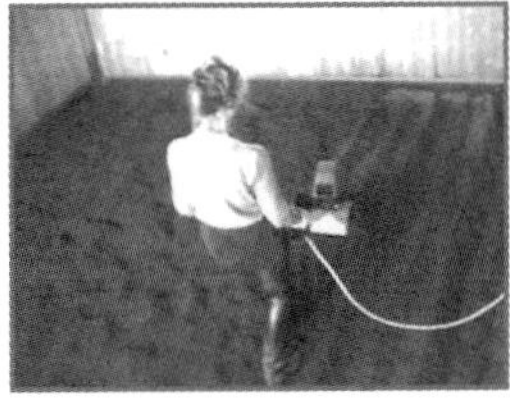

2. Mix the wet and dry Equidry together:

   Plug in the Equidry aerator. Run it through the stall bedding. The rotating tines churn the wet and dry bedding together and leave the bed neatly groomed.

   Time: 1.5 to 2 minutes

Total time:
3 to 5 minutes per stall.

Simply put, if urine works its way to the bottom of the 4-inch bed of Equidry (as it might if your horse urinates several times in the same spot, or if the Equidry is less than the optimal 4 inches deep), the bedding's immediate wicking capabilities can be overwhelmed. This allows urine to pool at the bottom. The mulching effect can hold it there, continuing to foil the Equidry's natural evaporative talents.

If this layer of mulched urine is allowed to remain long enough, it'll generate ammonia fumes. These fumes won't be noticeable until the bedding is disturbed enough to expose the wet underlayer, thereby interrupting the mulching effect. Then, the ammonia can hit you like a wall.

The best way to foil this factor is to disturb all the wet spots while they're fresh, during daily cleaning. I use the electric aerator available from the Equidry company—it's a blunted version of a lightweight electric rototiller. In less than 2 minutes the Equidry in my 17' x 20' stalls is churned into attractive, sweet smelling rows. This does two things: It mixes the wet with the dry, which ensures that every wet granule is in contact with a dry one, for rapid wicking and evaporation; and it aerates the bed, which encourages aerobic breakdown of the urine's components.

- Climate affects the rate at which urea breaks down and forms ammonia gas. The warmer the air, the faster this takes place. However, warmer weather can accelerate evaporation, helping Equidry do its job faster. Humidity is the deal breaker. If the temperature in the stall is warm and the humidity is low, evaporation of urine can occur faster than the production of ammonia gas, leaving no ammonia. Good ventilation can enhance this process. Poor ventilation can stymie it. If the air is moving in the stall in the "ammonia zone," between your ankles and your knees, and it's moving in an outward direction (taking it out and away from the stall), Equidry's evaporative skills are at their peak. Under these conditions, I've measured the ammonia levels in a heavily used Equidry-bedded stall. They're undetectable to my ammonia-sensitive nose—and to an instrument that registers between 2 ppm and 1,400 ppm. (For more on stall ventilation, see Chapter 5.)

- For horse owners living where temperature, humidity, and air movement aren't conducive to quick evaporation, a preemptive strike against ammonia is best. After a great deal of experimentation and several trials with so-called stall fresheners, it appears that the key is daily maintenance, rather than additives. Over time, I've figured out how to take fullest advantage of Equidry's evaporative strengths, avoid the mulching effect, and relegate ammonia fumes to distant memory. See "The Perfect Maintenance Routine," page 74.

Proof Positive

In one study, an Equidry-bedded stall that had been used for 6 months without vacuuming was allowed to "rest" for 12 days. Then the bedding was sampled and sent to the diagnostic lab at Washington State University's (WSU) veterinary school diagnostic laboratory, to be cultured for *coliforms*—the category of bacteria most often responsible for intestinal infections. The cultures consistently yielded no coliforms. In a subsequent study conducted at WSU's diagnostic lab, Equidry was purposely seeded with pure cultures of *Salmonella typhimurium*, a bug that's notorious for being highly contagious and persistent in a stable, putting any horses on the premises at risk of catching the disease. The contaminated Equidry was then heated to 115 degrees Fahrenheit for 4 hours, cooled, then cultured repeatedly. Even with nutrient broths they were unable to coax a single colony to grow. A follow-up study repeated the process, this time purposely contaminating the Equidry with a pure culture of *Streptococcus equi*, which causes strangles. Once again, the heat treatment left the bedding absolutely bug-free, and no amount of laboratory coaxing could revive a single colony.

Comfort

I've seen horsepeople pick up a handful of Equidry, adopt a thoughtful expression, then claim closed-mindedly that *their* horse won't like it, end of discussion. If they're thinking "soft and fluffy" is the only road to comfort, they're thinking like humans, not horses. Soft and fluffy, when you weigh over 1,000 pounds and urinate up to six gallons in your bed each night, is a rather absurd notion, but it's consistent with the belief system under which we've operated for generations.

Many horses that never slept in the recumbent position before will lie flat out on their side and snore on this bedding. This was the case with my imported foundation mare, Murkje* 9574, as of this writing 18 years old. I got her when she was 2 years old, and in the ensuing 16 years she'd lie down only when she was in labor.

The first night I put her on Equidry she slept on her sternum. I was delighted. The second night she went flat out on her side and snored. If it hadn't been for the snoring I would've thought she was dead. Now she sleeps like that every night, flat out on her side, for an hour at a time, for a total of about 5 hours deep sleep per night. I've never known her to be so comfortable. I made a video of it, snoring included, and sent it to the gentleman who invented Equidry, with thanks.

Though individual Equidry granules pinched between thumb and finger may seem hard and sharp, when you lie down on the material, it's supportive and relaxing, much like a bean-

figure 2.2
Friesian mare Murkje* 9574,
her second night on Equidry.
For 15 years, this mare never slept lying down. Now she sleeps this way every night.

bag chair. When my horses get up after taking a nap, there's an impression of their body in the bedding, just like that so-called "memory mattress" you've read about in your junk mail.

The bedding is cooler in the summer and warmer in the winter, because of its thermal mass and insulating properties. And, horses that are losing their coordination and strength because of old age or lameness—that is, those that ordinarily might be reluctant to lie down because they've had difficulty getting up again—appear to feel much more confident on Equidry, and more willing to lie down and get a good night's rest, because the hoof gains immediate and secure purchase in this bedding. There's no slipping. As a result, there are fewer hock sores. Horses that have foot soreness, who struggle across a gravel driveway, practically say "ahhhh" when they enter an Equidry-bedded stall, their stride lengthening and their limp becoming significantly less obvious.

Hoof health

Equidry shines in this regard. In my experience, horses that have had a long history of weak, crumbly hooves and flat, collapsed, tender soles improve vastly when bedded on Equidry. There's less flaring of the hoof wall, less breakage and chipping, less "powdery" hoof, better cupping (concavity) of the sole, and better resiliency of the hoof wall and sole—even in horses that spend the day grazing in a bog and only come in to the Equidry-bedded stall at night. I believe there are two reasons for this.

First, despite a lot of loudly expressed opinions to the contrary, a horse that spends the bulk of his time standing in a dry medium will have healthier hoof horn (see page 41). A dry environment simply makes for a stronger, more resilient hoof. Second, the conformability of Equidry shapes to the underside of the hoof and supports it vastly better than any of the other beddings I've tried (and I've tried all the ones that didn't have instantly worrisome characteristics). So, Equidry provides good foot support, a good cleaning out of the *sulci* (the crevices on either side of the frog), a daily foot massage, and exfoliation of dead frog.

In my two imported Friesian horses (who for more than 14 years had chronically flat, tender soles, flared hoof walls, and crumbly hoof horn with separation at the quarters despite diligent care), 6 weeks on Equidry bedding brought such improvement in their feet that it caught the surprised attention of my farrier, who'd attended these horses for 8 years. Six weeks after that first surprise, he was even more impressed. Six weeks after that, he ordered Equidry for his own stalls, and he is excited about the possibility of using it for the foundered horses he rehabilitates.

Remember those skin bumps?

Bedding my horses' stalls with Equidry is a good way to soothe and clear problem skin. The bedding harbors no biting or burrowing pests, no thorns, stickers, or slivers of wood. There's nothing in it to cause or exacerbate itchiness, making it the skin-safest place to be. The granules are mildly abrasive, so lying down on them tends to give the coat and skin a mild currying. In pasture, my horses dig in puddles and make mud, then lie down and roll in it. I know that in the morning all that mud will be in the Equidry. The good news is, when they walk out of their stalls they're dry, clean, and glossy.

Safety

Equidry bedding is absolutely fireproof. And, because it's a long-term bedding rather than a disposable one, it avoids not only the fire hazard of flammable bedding in the stall, but also the fire hazard of storing replacement bedding on the premises. It's also free of toxins and contains absolutely nothing that could be construed as allergenic, so it's safe for the environment and for your horse's respiratory tract.

Hygiene

Because Equidry is a long-term bedding rather than disposable, I worried whether hygiene might be an issue. With old-fashioned bedding, I could fool myself into thinking the stall was hygienic by stripping out the bedding, treating the floor's wet spots with a stall freshener, and adding fresh bedding. It felt like a clean thing to do. However, as I've already pointed out, most of the old-fashioned beddings harbor bacteria, molds, insects, mites, and spiders even before they've been put into the stall. And, if a horse were to come down with a contagious disease, disinfecting that old-fashioned stall would be a tall order. With Equidry, however, it's easy. The Equidry company is developing an electric blanket that heats the bedding to a little hotter than a scalding bath. I rake the Equidry into a mound in the center of the stall, lay the blanket over the mound, plug it in, and let it "cook" until a thermometer registers about 150° Fahrenheit throughout the bedding's depth, which it'll maintain—it doesn't get any hotter. Then I unplug, fold up the blanket, spread the bedding back out, and I'm finished.

Flies

From a fly's point of view, there's nothing attractive about Equidry. From a horse's point of view, that's attractive indeed. Flies are attracted to moisture, organic contamination, the smell of rotting urine, and manure. All of these factors are short-lived and easy to remove from Equidry. And, because there's no urine-soaked bedding in the manure pile outside, there are fewer flies on the premises to begin with.

Fire

When manufactured, Equidry is heated to 2000 degrees Fahrenheit. This achieves hardness, to make the granules virtually indestructible (only an estimated 1 percent of the product breaks down in a year of use). As a convenient side effect, this bedding has already been through the fires of Hades. It won't ignite, burn, smoke, or smolder.

Storage

There's no need for storage. Most of the Equidry I have in my stalls today will be here years from now. The company advises me to accept 10 percent loss from the stall per year. One percent or so could be lost to crushing, they say. Another 2 percent might spill out of the stall. Figure 5 or 6 percent lost when I clean out the manure, especially if the manure is on a urine spot. (Wet Equidry is more likely to stick to things, and to itself, and fail to fall through the tines of the manure fork.) Add another percent for a fudge factor, and that means I'll want to order a 10 percent load of Equidry every year (if I'm wasteful) or two (if I'm not), to top off my stalls.

Foals and Equidry: NOT RECOMMENDED!
I love Equidry in the foaling stall. When the "water breaks" there's no mess, it's infinitely cleaner than ordinary bedding if properly maintained, and when the foal makes his first attempts to stand, he can plant his feet in the Equidry and stand much sooner, without slipping.

HOWEVER, this is not to say that foals should be housed in stalls bedded in Equidry. It's simply too soon to say. Most beddings, including Equidry, alter the angle at which the foal's foot interacts with ground forces. For this reason, I have always advocated keeping foals outside, on the natural ground, rather than in a stall, no matter what you use for bedding. The interference with ground forces may or may not interfere with developmental forces in the foal's legs during his first formative weeks.

Until the research has been done, I recommend that you err on the side of caution and use Equidry (and, frankly, most any bedding) only for horses over six months of age. If you foal your mare indoors, on *any* bedding, I advise moving mare and foal to solid ground, preferably a firm, grassy meadow, as soon as the foal is standing. As the research is done and there are more answers than questions about this issue, I'll post the information on the Perfect Stall website: www.theperfectstall.com.

Cost and contact information:
To figure the cost of Equidry Bedding, multiply stall length x width, and multiply that figure by 5.55. So, a 12' x 12' stall would cost $800. plus shipping and installation (usually there's nothing to installing it). Call Equidry LLC at (800) 311-5767, or go online at www.equidry.com.

Why The Perfect Stall Loves Equidry Bedding

- Saves money on bedding, labor, and disposal.
- Spend 1/4 to 1/3 the time cleaning stalls, while keeping them cleaner and fresher-smelling than ever before.
- Less dusty, ammonia-free stalls, for allergy relief and improved respiratory health.
- Improves hoof horn health.
- Environmentally responsible.
- No wood-waste toxins.
- Fireproof.
- Significantly reduces fly population.
- Can be deep-cleaned and disinfected, without chemicals.
- No storage needs.

What would you do with an extra week-and-a-half per month?

Now that I have Equidry™ Bedding in my eight stalls, I
spend 2 hours per month mucking each one,
instead of 10 hours.
Factor in 20 minutes per month vacuuming each stall
(even though I do it less often than that), and my time
savings amount to 7½ hours per stall per month.

For eight stalls, that's a time savings of
60 hours per month.
That's 1½ work-weeks saved, each month.

What would you do with an extra 1½ weeks per month?

I'll tell you what I'm *not* doing with that time.
I'm not cleaning my house.
I'm riding my horses more. I'm taking Tai Chi lessons.
I'm taking daily walks with my husband.

I used to pay a high school kid $6 an hour to clean stalls.
It took her 2½ hours per day.
Now I'm saving $300 in wages, every month, or
$3,600 a year.

What would you do with that money?
I bought a new dressage saddle.
My next goal: an indoor arena.
Paid for with money I used to dump on the manure pile.

Think of me the next time you're mucking stalls.
I'll be riding. Or practicing white-crane-spreads-its-wings.
Or breathing the fresh air on the trails.

Figure 3.1
Feeding, the natural way.

Chapter 3
Feeder

What's wrong with my feeder?
What's the best way to serve dinner to your horse in his Perfect Stall? Let's start with how horses were designed to eat.

In the wild, horses eat pretty much constantly, with a lowered head. You might think the reason for this is obvious—horses aren't giraffes; they're grazers, not browsers. The leafy grass, fresh grains, and seeds they eat are down low, at fetlock level.

But there are other reasons for this posture, as well.

For example, your horse's long-distance eyesight (for watching the horizon for predators) is optimized when his head is down and his eyeballs are rotated up to the horizontal. In this position, his eyelids cover over more than half of the orbs, and his eyelashes extend outward like protective awnings, to keep chaff and seeds and dust from settling on his eyes as he grazes.

And, his long neck and its two pipes—the esophagus and the *trachea* (windpipe)—are angled downhill. That way, the feed he picks up starts at the front of his mouth and is worked to the back by his tongue, against gravity. The feed lingers longer in his mouth as a result, getting pulverized more completely by the grinder teeth at the back.

This breaks up his feed into smaller, easier-to-digest pieces, and also stimulates the production of more saliva. Saliva starts the digestion process (thereby avoiding indigestion and excess gas), helps neutralize excessive stomach acid (thereby avoiding ulcers), and coats the bolus of chewed feed with a slimy layer of lubrication. As a result, the feed bolus slides more easily through his esophagus, through which it must be conveyed *uphill*, against gravity, via muscular contractions called *peristaltic action*. Any bolus that's too large, inadequately chewed, or poorly lubricated is likely to fail the swallowing process and fall back into his mouth for further processing. As a result, choke is averted.

While your horse's head is lowered, the valve-like opening into his stomach, at the base of the esophagus, remains folded closed until a bolus of feed is delivered, which causes that muscular "lid" to open. It closes itself as soon as the

bolus passes through. All this ensures that when your horse swallows a bolus of chewed feed, its size will be manageable, it'll be slippery and free of sharp, irritating edges, and, once it's in the stomach, it and other stomach contents aren't likely to come back up.

No belching

This is why horses have a reputation for never belching—the "lid" at the top of the stomach is open only during the last part of a swallow, if his neck is angled down. When your horse's neck is elevated, however, the "lid" spends more time in the open position, which makes the base of the esophagus more vulnerable to being damaged by partially regurgitated stomach acid. Resultant scar tissue can gradually narrow the esophagus and damage the valve, leading to chronic choke (a tendency for feed to get stuck in the throat) later in life.

Clean lungs

Your horse's lungs and *airways* (the tubes that carry air to and from his lungs) need to clean themselves out constantly, or they'll accumulate debris and get infected. That debris includes stuff he's inadvertently inhaled, such as dust, pollen, molds, chaff, bugs, and the like. It also includes dead cells his respiratory tract naturally sheds, plus natural mucus and bacteria.

In the wild, all that stuff drains out while he's grazing, head down. It's an efficient design for the wild horse because he spends so much of each day—roughly 21 of every 24 hours—with his head down, grazing. The more time your horse spends with his head elevated, the greater is his risk of getting *pleuropneumonia*, an often deadly lung infection. That's because all that normal debris dribbles down into his lungs, instead of draining out. This is why a horse in a trailer, head tied up short, is at risk for getting "shipping fever" (pneumonia or pleuropneumonia). The normal respiratory debris, including bacteria that normally inhabit his mouth, nose, and throat, dribble down into his lungs instead of out onto the ground.

So now you know why horses need to eat with their heads lowered. It's better for their

- eyes,
- throat,
- digestive tract, and
- respiratory tract.

And how are most horses fed? You guessed it. *Heads up.*

Most stalls have an overhead feeder, with a rack for hay and a tray to hold grain and catch *hay fines* (dust, leaves, and short pieces of stem that fall from the hay rack). The design is clever and seems to be convenient in some ways. However, in a moment I'll explain why it's really not what I wanted for The Perfect Stall, in addition to the fact that it's not what's best for your horse.

If given the choice, your horse probably eats his grain first. If the hay's already in the feeder, his eyes are buried in hay that poke through the bars while he's gobbling up his grain. When he finishes the grain and begins pulling hay from the feeder, he must hold his head raised and extended. This naturally rotates his eyeballs downward, exposing the entire iris (the colored part of his eye). Plus, in this position, his eyelashes no longer stand between his eyeballs and the chaff, stems, and dust. As a result, eye irritation is daily fare, and he's more likely to damage his eyes further by rubbing. Scratched corneas, pink-eye, and traumatized eyelids are common in horses fed overhead. And, the more "water" that runs from the irritated eyes, the more flies are attracted to them. Hay fines get in his forelock and in ears, further encouraging him to rub, and inviting self-injury (such as abrasions and lacerations to eyelids, face, and ears), as well as a rubbed-out forelock and mane.

figure 3.2
The overhead feeder:
bad for the eyes,
bad for the lungs.

Most stalled horses are even more predisposed for inadequately chewed, inadequately lubricated mouthfuls to back up, because they're fed their entire daily ration in two huge meals, instead of in small mouthfuls throughout the day. This, despite having a stomach that can accommodate only about 2 gallons total (and that's a stretch). We've already discussed the fact that horses fed headsup-style often belch audibly, suggesting that their stomach "lid" is open while their heads are raised (see page 91). It's logical that they're more prone to esophageal damage from refluxed stomach acid this way, and therefore more prone to chronic choke because of shrinking scar tissue constricting the esophagus. Respiratory infections, allergies, chronic cough, and decreased performance all are more likely when fed high.

There's another problem with the overhead feeder, relative to maintenance of The Perfect Stall. It's messier. More hay and dust fall onto the bedding. Not a problem when your horse's stall is bedded in hay, straw, or sawdust, because those beddings are already messy and dusty. But as you know from Chapter 2, if you're using the Perfect bedding, the less hay that spills, the better, because spilled hay and fines get mixed in with the bedding and can cause an increase in dust and ammonia fumes.

The Perfect feeder

The feeder I chose for the Perfect Stall is the Pro-Panel Corner Feeder, designed by Washington horseman Pat McCarty. It sits on the stall floor, in a corner, fastened to the walls with a lag screw. (I used a couple 2"x 4" pieces of plywood as makeshift washers to prevent the screw head from pulling through the plastic.) Installed in this way, your horse won't be able to drag the feeder out of the corner or knock it over.

figure 3.3
The Pro-Panel Corner Feeder.

Chapter 3 Feeder

The Pro-Panel corner feeder is made of durable, high-impact plastic, molded without seams. It has a deep central bin for hay, flanked by two smaller, higher chambers for salt and grain. All edges are rolled; there's nothing worth rubbing on. And, because the feeder stands tall enough to reach approximately chest-high on most horses, even a horse that likes to paw is unlikely to get a foot in. And, if he does get a foot in, it's easily retrieved, with little risk of injury. The plastic is heavy, resilient, and very strong, so my 1,600-pound, 17.1-hand pawmaster causes no damage to himself or the feeder when he strikes the thing with his front feet.

While I awaited the arrival of my Pro-Panel feeders, I used empty 50-gallon water tubs for my horse's hay and grain. Although they were an improvement over the usual heads-up feeders, both for the horses and for keeping the stalls clean, one mare consistently flipped the hay out with her head while chasing after the last few grains and leaves. With the Pro-Panel feeder, her grain is separate from her hay, and her hay stays put in the narrower, deeper hay chamber which discourages hay-flipping.

I stand a trace-mineralized salt brick in one side dish and pour plain loose salt all around it, so my horses can choose one or the other at will. I put grain in the other side. In the morning, my horses' faces, forelocks, and ears are no longer full of hay chaff. Their eyes are clean and clear, and there's little if any hay spilled onto the bedding, which means less waste, and less work for me. They snort less, which tells me their respiratory tracts are draining properly and they're getting less dust up their noses.

The Pro-Panel comes in two sizes: small (which comes in a variety of colors) and large (black only). The main advantage of the small feeder is that it can be delivered by UPS; the larger one is shipped by truck from the manufacturer, directly to you or to your local farm store.

However, in my opinion, any horse of average size or larger will need the large feeder. The side chambers in the small Pro-Panel feeder look too pinched to comfortably accommodate the average horse's muzzle. To be precise, the average muzzle could fit in, but the horse might not be able to move his lips around sufficiently to pick up feed without frustration. And, the small feeder is quite a bit shorter, which could invite a pawing horse to stick a foot in and foul the hay. I use the smaller feeder for foals, in their creep corral. But for grown-up, average-size horses up to warmblood category, the large feeder is the way to go.

For durability and safety reasons, the top front edge of the Pro-Panel feeder is rolled inward. This reinforces the edge and also discourages using the feeder for scratching.

I do clean it out daily, by lifting out any hay left behind, and by wiping out any dust, dirt, or chaff with a damp cloth, which also wipes up excess saliva. I keep a rechargeable hand-held vacuum in the barn for cleaning excessive dust and dirt from the feeder bottom. I also keep handy a box of clean rags, a bucket for used rags destined for the laundry, and a 1-quart spray bottle of water mixed with a teaspoon of white vinegar, for a nontoxic cleanser.

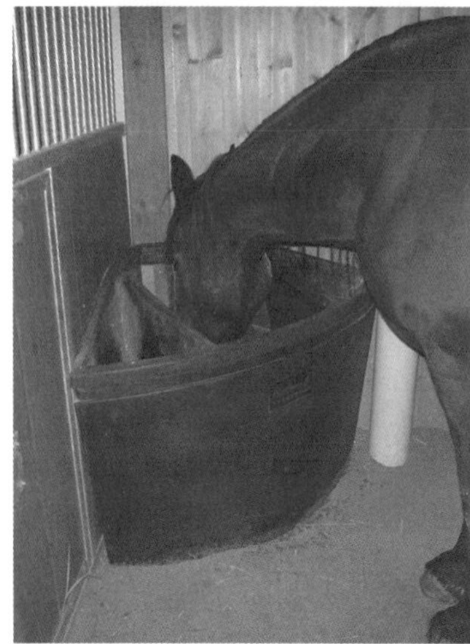

figure 3.4
The feeder in use.

Clean leftover hay out of the feeder each day. Your horse will leave hay for one of two reasons: either he's not hungry enough to eat it, or there's something wrong with it. Either way, it's important to know. The grain compartment should be cleaned daily too—if mice get into it, they're likely to leave deposits. If the stall is maintained by somebody other than you, make sure they know to do this, and be sure to check that they're complying.

Improving perfection

To make the Pro-Panel Corner Feeder even better: Install a removable, perforated, false floor, to hold the hay about 2 inches off the bottom of the feeder. You can even vent the chamber beneath the floor to the outside by drilling a hole in the plastic, inserting a 1" diameter PVC pipe, and installing a small, solar-powered exhaust fan, to pull fine dust down and away from your horse's nostrils while he eats. If your horse has a respiratory disorder, such as heaves or allergies, everything you can do to minimize dust will help him.

Cost and contact information:

The large Pro-Panel feeder costs $199 plus shipping and is available directly from the manufacturer at (208) 989-9991, or go online at www.propanel.com.

Why The Perfect Stall Loves Pro-Panel Corner Feeders

- Encourages your horse's respiratory tract to drain properly, keeping it cleaner and healthier.
- No hay chaff, dust, and other debris falling onto your horse's eyes and eyelashes.
- Helps keep stall bedding cleaner by reducing hay spillage.
- Durable, high-impact, no welds to pop, no metal bars to bend.

Chapter 4
Water

They say water is the most important nutrient. That's because none of life's processes can take place without it. No matter how healthy, athletic, vibrant, thin, fat, tough, or gorgeous your horse may be, he'll be in big trouble within 24 hours, and can perish in a few days, if he doesn't have access to drinkable water.

Your horse drinks 8 to 15 gallons per 12-hour period. He needs about 50 percent more than that available to him, because he's somewhat wasteful. A number of factors—exercise, hot weather, lactation, fever, chronic disease, dry feed, excessive urination, and more—can increase his average water intake by as much as threefold. When he's confined, he can only drink what you give him. It's important to give him more than he needs in an average day, in case his needs on any particular day are higher than average.

For horses that spend at least part of every day in a stall, there are two options for providing water: some sort of tub or bucket, or an automatic waterer—one of several types of devices that refill themselves from a water source plumbed right into the drinking vessel. There are potential problems in both options.

What's wrong with my waterer?

Tubs or buckets get knocked over, defecated in, slobbered in, or cracked. Horses that are bored—and horses that have a sore foot or leg—have been known to put one or more feet in the water, and of course they won't drink it after that. Mice and other critters in search of a drink commonly fall into buckets and drown, and I have yet to meet a horse that'll drink around a floating corpse.

The water in the container can freeze in cold weather unless steps are taken to insulate and/or warm it, which raises the risk that the horse will chew and eat the insulation, or chomp down on, step on, or get tangled in the electric cord. Though submersible and floating water heaters abound, the obvious hazards of mixing water with electricity make such heaters a gambler's insurance policy—you may not be able to see the fine crack in the heating element, but the shock it delivers your horse when his mouth touches the water will put a serious crimp in his willingness to try again, if it doesn't kill him outright. You could get shocked, too.

Chapter 4 Waterer

Your horse wants fresh, clean water and will shun a water source that's anything less. He'll also rinse his mouth and muzzle in the water just before drinking, making it less and less "drinkable" with every swish. Have you ever noticed there's almost always a gallon left in the 5-gallon water bucket? Your horse won't drink that last gallon because it's got too much slobber in it.

After years of dipping and carrying four full, sloshing, 5-gallon buckets of water from a cistern 30 feet from my barn to each stall twice a day for 15 years in all kinds of weather, I thought I'd died and gone to heaven the day a contractor installed a spigot in the aisleway. I gleefully got a hose long enough to reach every stall and started making plans for all the time and muscle power I was going to save.

What I hadn't considered was that keeping the stall buckets or tubs filled is only half the job. The other half is *emptying* what's left and cleaning the vessel before the fillup. Not such a big deal when you're 18 years old, but there comes a time when you'd like the option of getting your horse chores done without breaking a sweat and throwing your lower back out.

And then one day, a dream came true—or so I thought. I got my first automatic waterer, plumbed into a developed spring and installed so that horses from two adjacent pastures could help themselves to fresh spring water whenever they wanted it. This was going to be *sooo* much easier. The waterer refilled itself whenever an internal float valve reached a pre-set low point. Once the horses learned to press on the floating disc (which looks like a bloated Frisbee®) that covered the water, I had it made. Or not.

A week later, I discovered my first disappointment. This thing was impossible to truly clean. Designed sort of like a toilet bowl and tank, its drinking reservoirs intercommunicated with a central, mostly inaccessible inner section full of nooks, crannies, corners, delicate looking gadgets, and obstacles. The floating discs over the watering holes blockaded my scrub brush, so I had to carry a socket wrench and two different size sockets to get to the slime.

I quickly learned to lower my standards of cleanliness, or risk popping a vein over all the slime-coated areas I couldn't reach, even with the top off. In fact, the only way to get all the surfaces clean would be to carry a pocket full of Q-tips® and a toothbrush along with my socket wrench. Or, I could pour bleach in the thing every week. That would cut down on the germ population in the water, but it'd also cut down on the amount of water my horses would drink.

To make matters even dicier, the two drinking reservoirs were connected, meaning that water from one side was free to mix with water from the other. That meant I had no hope of preventing germs from one pasture group spreading to the other group, though the germs might stop in the middle chamber to inoculate the slime layer for future fun.

Two months later, the first heavy frost of the winter came, and I plugged in the submersible heater. I remember feeling disappointed that I was still exposing my horses to the risk of electrocution, even after spending the $225 to buy (and $150 to install) what I thought was a state-of-the-art waterer.

The following week something went wrong with the float that regulated the water level (disappointment number three), and I had one heck of a flood, which emptied the spring and created a 50-foot skating rink around the waterer. The only silver lining I could think of at the time was, it's a good thing this didn't happen inside the barn, and that none of my horses fell on the ice and broke something. I contacted the shop that sold me the waterer, thinking optimistically that I might be able to get a new one. I was young and foolish.

Then, one Saturday night in mid-February, the heating element died. Strike four. How did I know it died? The waterer froze solid. My husband and I set a space heater beside the enormous ice sculpture that encased the waterer, and draped a tarp over the whole thing. We waited several hours for the ice to melt and reminded each other how glad we were that we were in the habit of checking the waterer every day; otherwise, our horses might have gotten dehydrated and colicked.

Even when they don't malfunction, there's another potential problem with automatic waterers: You can't tell how much water your horse is drinking.

That's no big deal if your horse is healthy, but what if there's a problem brewing, and the only way to catch it early is to notice that he hasn't been drinking? With most automatic waterers, the only way you can tell if your horse is drinking enough is by checking to see if his manure is dry and/or he's clinically dehydrated, and by that time it's a little late.

I have one additional problem with this first waterer of mine. It sits low to the ground, and my horses have taken to dunking their feet in it. This isn't conducive to having drinkable water available, and I fear that the molded plastic housing won't stand up to the abuse forever. There'll be a flood when it breaks down, but part of me will rejoice, because then I'll be able to replace that waterer with a Perfect one.

figure 4.1
The Nelson model 730-10W.

The Perfect waterer

You probably won't be surprised to learn that I chose a Nelson waterer for my Perfect Stall. This is the sleek, shiny, expensive-looking waterer you've seen at high-end facilities. As Nelson reps are quick to point out, every zoo in North America has Nelson waterers—there are 3,000 of them at the San Diego Zoo alone—and all but one or two of the nation's veterinary schools have them, as well.

What's so good about Nelson waterers? It'd be easier to tell you the one "bad" thing, which put me off right away: the price. Figure on about $380 for a stall (one-horse) model like mine (see figure 4.1 on page 106), plus installation. Ouch, I thought.

But hold on. For my troublesome waterer described above, I paid $225 up front, $150 for installation, plus about $75 worth of replacement parts when the blasted thing let me down not once, not twice, but three times. That's $425 real out-of-pocket money, plus a whole lot of aggravation. In retrospect, would I have been better off with a top-of-the-line, serious watering device that's better designed, more reliable, and safer? Stupid question. Here's what I figured, in retrospect.

1. The Nelson has a safe heating system for wintertime watering. In fact, it's the only livestock drinking water heating system that's UL (Underwriters' Laboratory) approved, meeting all builders' codes for all barns in all states. No electric heating element touches the water itself. It's a brilliant design. The element heats the air *under* the stainless-steel watering bowl, governed by a thermostat that's factory set to keep the water at a steady 50 degrees Fahrenheit. That's the natural temperature of groundwater below the frostline and significantly warmer than the 32-degree water a horse would drink from an ice-encrusted bucket. It's a proven fact that horses drink more water during cold weather if the water isn't so doggoned cold.

2. There are no float valves to get stuck and cause a flood. Instead, filling is activated by the weight of the bowl (or, to be precise, its weight *loss,* as water is consumed). When the bowl is comfortably full, its weight shuts off the spigot.
3. Because the wide drinking bowl has a relatively small (1- quart) capacity, it refills every time a horse takes a drink. That means very little water is wasted, and my horses have *really* clean, aerated water in front of them, all the time, as if there were a babbling brook running through their stalls. This is a tremendous health advantage—the cleaner the water, the more a horse will drink. The water bowl is completely separate from the inner works of the waterer, which makes thorough cleaning a breeze. In fact, the water bowl lifts out for daily cleaning (no tools required). Simply dump it out, wipe with a soft cloth (or, if desired, run it through your dishwasher), and put it back in place. Fresh water immediately runs into the clean bowl, and you're done.
4. When more than one waterer is supplied by the same pipe, Nelson's USDA-approved non-siphoning valves keep the waterers from intercommunicating—in other words, none of the water from a sick horse's bowl will enter any other waterer.
5. The Nelson's all-stainless-steel construction makes it durable and attractive. It's round and smooth, has no sharp edges and no corners to catch on. When installed as recommended, the drinking bowl sits at chest level

for the average horse. This places it within easy reach of thirsty mouths, but out of easy reach of dirty feet.

6. Daily water intake is no mystery with my Nelson waterer, because I ordered the consumption monitor. It's an electronic, LED-display timer that's plugged into the water-supply pipe for each individual waterer. Rather than measuring the actual amount of water the horse drinks, it measures the amount of time the water ran since the last time you re-set the monitor. This gives you all you need for assessing whether your horse is drinking less than he usually does.

Cost and contact information:
The 730-10W model cost about $380. Call Nelson Manufacturing at (888) 844-6606 or go online at www.nelsonmfg.com.

Why The Perfect Stall Loves the Nelson Waterer

- Reliable, fresh water on demand. Less concentration of slobber. Water-use monitors.
- Free-flowing in freezing weather without risk of electric shock. Holds water at steady 50°F in winter, when horses will drink more if the water is not frigid.
- Sanitary. Low maintenance. Easy to clean and disinfect.
- No intercommunication between connected waterers.
- No floods from stuck float valves.

In case you didn't know...

Good ventilation is your horse's best protection against lung disease, allergic conditions such as heaves, and respiratory tract inflammation.

Chapter 5
Ventilation

What's wrong with my ventilation?

Most older barns don't have a ventilation system per se. They have windows, and if the barn manager is savvy, all the windows that can possibly be left open are left open. Cross-ventilation is the mantra of those who think about such things.

Newer, fancier barns often have a bona fide ventilation system designed right into the blueprints. This design usually involves an enormous, attic-style electric fan roaring in the cupola, and data to prove how many air exchanges per hour the system provides. But even these systems aren't doing the job, at least not from the point of view of your horse's lungs. I've stood in stalls that have ammonia fumes so strong that eyes sting and tears run, while the barn owners and managers adamantly claim they have a "top notch" ventilation system.

Here's the problem. A well-ventilated barn is one thing. A well-ventilated *stall* is quite another. The ventilation system might be top notch, but it's not likely to have much of an impact on the air behind the stall doors.

As a result, the people walking in the aisles are comfortable, while the horses' lungs are marinating in respiratory irritants.

Here's why those expensive ventilation systems fall short. Even if there's a window in the stall, it's likely too high to impact where the ammonia fumes tend to hover: between the floor and the horse's knees. I've proven this, using a beekeeper's smoker. The smoke curls around for a moment in midair, then settles in the ankles-to-knees zone, and hovers there.

This happens even in fancy "well ventilated" barns, and in musty old converted dairy barns like mine. If the stall door is closed, the smoke settles and stays there.

Clearly, your horse's lungs deserve cleaner air. But what's the answer, if whole-barn ventilation isn't doing the trick? What about a box fan?

No, that doesn't work either. It stirs up the smoke, but it doesn't move it out. No matter how the box fan is aimed, the volume of air moved is insignificant. I didn't believe this at

figure 5.1
A $40 ceiling fan, priceless stall ventilation.

first, because when I walk in front of a box fan, the blast of air is downright unpleasant, like a harsh draft, so it seemed like it was moving *lots* of air. But the goal is not a narrow blast of air in hot weather; the goal is *ventilation for the whole stall year-round*, for the purpose of fresher air, cleaner lungs, and helping our Perfect bedding perform to its evaporative best. In my early experience with Equidry Bedding, before I started churning the wet spots into the dry (see page 74), I tried to enhance evaporation by aiming box fans at the floor. They were worthless.

Perfect ventilation

The solution is so simple, it makes me feel really stupid that it took me years to figure it out. A ceiling fan. One per stall.

I went to The Home Depot and bought a $40 model for each stall. No frills, no light; just a fan. The particular brand I bought was Hampton Bay, but there are other brands with similar features, available at hardware stores, discount department stores, and home improvement stores. Get one that's UL-rated (approved by Underwriters' Laboratories) for indoor or outdoor use, is reversible, and has three speeds. I had a certified electrician wire mine in, then I flipped the switch, and set it to blow down, onto the stall floor, at high speed. I've left it on ever since.

The results are nothing short of miraculous. There's no harsh draft, only constant air movement that caresses the entire stall. Now my beekeeper's smoker reveals an entirely different scene: the smoke goes down in the center of the stall, crosses the floor in all directions, and disappears. And, when I purposely wet down the entire Equidry bed with the garden hose just to see what the moisture does, the bedding is completely dry within an hour. Ammonia fumes? They don't stand a chance with the combination of the aerator (page 74) and the ceiling fan.

Cost and contact information:
Check the lighting department at any warehouse home-improvement store. At The Home Depot and Lowe's stores locally, I found two brands of ceiling fan that are plain, Underwriters' Laboratories-approved for outdoor use (there'll be a UL on the box), reversible, and with 2 or 3 speeds, for around $40.[oo]. Make sure your stall ceilings are high enough to

accommodate a ceiling fan, then have your fan(s) hard-wired by a certified electrician.

If you want to save on electrician costs and combine your ceiling fan with your Perfect Stall light (see Chapter 6), and if your stall ceiling is high enough to accommodate the two together, choose a ceiling fan that has a standard bulb socket, or one that will accept a single bulb "light kit" (available in the same department for about $3).

Why The Perfect Stall Loves Ceiling Fans

- A fraction of the cost of whole-barn ventilation.

- Better respiratory health, thanks to a constant supply of fresh air in the stall, where it matters most, without a harsh draft, while pushing out stale air, dust, & ammonia fumes.

- Less risk of electrical shorting compared to box fans, particularly if hard-wired; no extension cords.

- Significantly reduce horse's exposure to weak-flying insect pests such as *Culicoides* gnats, and reduce fly population by assisting in evaporation of urine from Equidry Bedding, before ammonia emissions form.

In case you didn't know...

Incandescent light bulbs are a common cause of barn fires. Old wiring and cheap sockets meant for 60 watt bulbs get hot when you try to improve lighting by using 100 watt bulbs. And, the bulbs themselves are hot, fragile, and prone to breakage and shorting when used in barn conditions, with wide temperature and moisture fluctuations.

Chapter 6
Lighting

What's wrong with my lighting?

As a veterinarian, I consider poor stall lighting to be one of my biggest headaches. Lights tend to be abundant in aisleways, feed rooms, and grooming areas, but inadequate in the stalls themselves. Stalls that do have light usually have only one fixture. In basic barns, it's usually a bare bulb which not only casts deep shadows that obscure a horse's crevices, but also glares directly into my eyes. Hitting a vein to take a blood sample or give a shot is tough when the lighting is poor. Even finding a horse's anus to insert a thermometer is a challenge. If your veterinarian needs to conduct an examination, or block a nerve, or set a stitch, it'd be nice if your horse didn't have to be led out of the stall in search of better lighting, especially if there are medical reasons not to move him.

Chapter 6 Lighting

Good lighting in your horse's stall may not seem like an important feature until something goes wrong. If your horse is sick or suffers an injury, good lighting is essential for accurate assessment of his condition—your horse can't tell you how he feels, so most of your ability to protect his well being and manage his care rests on your powers of observation. You need good light to check whether your horse's gums are a normal pink, and to see whether the "whites" of his eyes are truly white or have a sickly tinge. In poor lighting you might not notice that your horse didn't clean up his feed because the leftovers are in shadow. You can't tell whether his manure looks normal, or if the quantity is right. You likely won't spot a problem until it's obvious your horse is in trouble.

Safety issues

There's more to good lighting than good quality light. There's also the matter of safety. The need for good lighting presents a bit of a conundrum, because there's so much about horse barns that constitutes a fire hazard: electric lights, which can short out, and hot bulbs, in a building that's chock full of flammable materials. I routinely see bare, 200-watt bulbs coated in flammable dust and brown dots of dried fly spit, and draped in cobwebs laced with bits of old hay, dead moths and hair.

The fire hazards in the average horse barn are like dead cars in the back forty—after a while people grow accustomed to their being there, and cease to notice. Lights may be installed low enough that if your horse were to rear up for any reason, he could hit his head and break the bulb. Power surges are another cause of bulb failure and can lead to spontaneous breakage. If lights are installed beneath a hay loft, concussion from dropped bales on the floor overhead can fracture fragile filaments and shatter the bulb. Metal cages can help protect a bulb from a head butt, but they don't protect your horse from falling glass if breakage occurs for any reason, nor do they keep flammable materials from touching a hot bulb.

Incandescent lights normally produce a lot of heat, and indoor-rated bulbs weren't meant to be used in unheated, damp facilities. As a result, they're exposed to wide variations in temperature—searing heat inside the bulb when the light is on, and cold winter temperatures and excessive moisture outside the bulb—which can shatter the glass. When the fragile filament inside a bulb starts to wear out, switching on the power often causes a small explosion inside the bulb, fracturing it. As shards of glass rain down in the stall, the risk of fire climbs up another notch because of electricity continuing to surge through the broken bulb. Moisture, even vapor, can cause corrosion and electrical shorts, pushing your barn another notch up the fire-hazard scale.

If you run a horse breeding operation, you might be inclined to use lighting to bring your broodmare into season sooner. If you show your horse, you might use lighting to prevent the growth of a woolly winter coat. Trouble is, if you install bulbs high enough overhead to be safe, you'll need a higher wattage bulb, and/or more than one fixture per stall, for sufficient foot-candles of light to affect your horse's inner sense of season.

Put a higher wattage bulb in that 60-watt socket and you're really ramping up the fire risk. You're also burning up valuable energy in a most inefficient manner. It's environmentally irresponsible, not to mention hard on your checkbook.

The Perfect Stall light

I chose light fixtures from Orion West Lighting for The Perfect Stall. Orion West proudly identifies itself as the "agricultural lighting specialist," featuring cool fluorescent light fixtures designed for barn use, and *equine* barn use in particular.

The company's tough, commercial-grade fixtures are fully encased in a durable, rust-proof, gasket-sealed fiberglass housing that keeps bulbs confined and isolated from dust, vapor, bugs, and debris. They're UL-rated safe even in wash racks and other wet locations, including outside under an overhang. Although I haven't tried it, the Orion West literature indicates that its OW-model lamp housings are so waterproof and tough that they can be power-washed without fear of breakage or leakage.

If you've ever accidentally switched off a mercury vapor light and had to wait 20 minutes to turn it back on, you'll appreciate Orion West's instant-on and full performance even in winter temperatures to –20° F.

The unit's fluroescent bulbs are long-lived and less fragile than most incandescent bulbs, therefore significantly less likely to fracture. They're bright enough that they can be installed well

figure 6.1
Orion West's fixtures.

Here's an idea...
If you want to save on electrician costs and combine your Perfect stall light with your ceiling fan (see Chapter 5), and if your stall ceiling is high enough to accommodate the two together, choose a ceiling fan that has a standard bulb socket, or one that will accept a single bulb "light kit" (available in the same department for about $3). Then, simply screw your Orion West light into the socket.

out of reach of High-Ho Silver's head at the peak of his most spectacular pose, while still providing more than enough light to affect his haircoat and a mare's reproductive season in the average-size stall. Because they're fluorescent lights, they're cool to the touch, which makes them fire-hazard-free. They also save you 80 percent in energy costs.

For stall use, there are two models to consider: The smaller, disc-shaped OW series, and the narrow rectangular AG series. I have one of each in The Perfect Stall: the AG over stall center for serious working light, and the dimmable OW towards the front of the stall. I use a dimmer switch to dim down the OW model to 50 percent when I'm night-monitoring a horse (for example, when on foaling watch).

The OW model lamp, about the size and shape of a layer cake, uses 36 watts to produce light equivalent to 200 incandescent watts, without glare or shadows, and with nearly full-spectrum light for true-to-life colors. The light produced, when hung 10 feet over the center of a 12 foot x 12 foot stall, is sufficient for breeding-season manipulation and winter coat suppression.

In retrospect, One OW-model lamp would have been plenty for The Perfect Stall. It comes in three styles: hard-wired to a junction box, retro-fitted to a "jelly jar" light fixture, or screwed in a standard light socket. All three are fully self-contained and dimmable, with bulb, diffuser, and housing in one replaceable, 10,000-hour unit.

The AG 9000 model, a narrow, 4-foot long fluorescent fixture, holds three nearly-full-spectrum tubes in a 7½-inch wide fiberglass housing, Using only 100 watts'-worth of energy, it provides light equivalent to 750 incandescent watts. It's amazingly bright, not at all harsh, bathing the stall in natural-looking light with true color perception.

Cost and contact information:
My screw-in OW-model light cost $52.95. Call Orion West equine specialty lighting at (877) 351-0841, or go online at www.equinelighting.com.

Why The Perfect Stall Loves Orion West's
Equine Lighting

- Significantly less energy for significantly more light.
- Improved visiblity and color perception, for accurate observation of your horses' condition, and easier veterinary exams and treatment.
- A single OW-model lamp produces sufficient intensity to induce early estrous cycling and help prevent winter coating.
- Safer: cool, fully contained, watertight and sealed.

In case you didn't know...

Research has shown that the risk of colic increases significantly when a horse's caretaker is somebody other than his owner.
Researchers speculate that this is because nobody knows your horse better, or watches him more closely, than you do.

Chapter 7
Surveillance

Your peace of mind as a horse owner is knowing, day and night, that your horse or horses are safe and sound.

If you're supremely lucky, you've got someone to help you who's as observant and horsecare-savvy as you are, and who's happy to keep a close eye on things so you can get away once in a while.

If you're like most of us, however, the only way to know for sure your horses are okay is to eyeball them yourself. The bottom line is that although you adore your horses and spend as much time with them as you possibly can, you also have a life outside the barn. As a result, it feels like you have only two options:

1. Abandon your “other” life, or
2. Worry constantly whenever you’re away (and hurry back).

It’s a setup for frustration, especially if you are, like me, wedded to someone who’s a horseman by marriage only. In theory, a stall camera is the perfect solution.

In the not-so-distant past, stall cams were a luxury affordable only for the wealthy. And, the cameras themselves were touchy, problematic, a chore to install, and available only from high-tech security companies. Most units required wiring between the barn and the viewing area. Those few that were wireless were limited to 100 feet between camera and receiver and had lousy reception, for flippy, snowy, indistinct images in black-and-white only.

Metal barns and barns with metal roofs confounded the unit’s ability to transmit and send a clear signal from barn to house. To make matters worse, because the barn environment is sometimes hot, sometimes cold, sometimes wet, and often dusty, the only closed-circuit TV cameras that could hold up were the most expensive models.

Times have changed, and so has the technology. A stall cam is now feasible and affordable for the home stable as well as for the boarded horse in a rented stall. There are wireless models sold by (and for) horsemen, designed to accommodate metal barns. They’re widely available and affordable.

And, if you have high-speed Internet uploading, you can ask your service provider or webmaster to hook you into a live webcam system. That way, you can check on your horse while at work, school, shopping, visiting, or on vacation. All you need is a computer and Internet access.

The Perfect stall cam

For the Perfect Stall, I chose Saddlebrook's BarnCam™, specifically their wireless system that's made for metal barns. With this system, I get full-color, live images and sound from each of four cameras, transmitted to a four-channel receiver plugged into the wide-screen TV in my living room.

My house is 300 feet away from the barn, and the system is rated for "up to 500 feet." For an additional sum I can add a high-gain receiver and double that distance, to 1,000 feet.

The receiver is controlled by a remote (included), as well as by a built-in timer that can be set to switch automatically from camera to camera at an interval of my choice. The remote and automatic switcher may seem like excessive luxury; however, I can now actually get some sleep on the couch when, for example, I have a sick horse. I'm not a sound sleeper when concerned about a horse, and with this setup I can turn up the volume and catch a little shut-eye, secure in the knowledge that if a colicky horse takes a turn for the worse, I'll hear it. If it turns out he's just scratching his back, I can tell by opening one eye and looking at the TV screen.

A Radio Shack device called an Audio-Video Sender (about $100) relays the signal wirelessly to other monitors in the house. If I'm not watching a sick horse, I may opt to move the surveillance post from living room to bedroom.

If you have a metal barn, you can separate the camera from the transmitter, then strategically position the transmitter to avoid signal interference from metal walls or roof overhangs. There's about 60 feet of wire between the camera and the transmitter, but if you need extra wire, Saddlebrook will provide it at no extra cost.

The camera's wide-angle lens, which causes very little birdseye-type distortion in the image, gives a much wider view than other cameras I tried. When mounted to the rafters over the stall door, one camera gives me an intimate view of the entire 17 foot x 20 foot stall—no blind corners.

Until just before press time, I had only two minor complaints about the Saddlebrook system, and they've been remedied.

First, unlike the camera and transmitter, which can be mounted apart from each other, the receiver's tiny paddle-like antenna was permanently attached to the base. I was unable to get a clear signal through my living room walls, but I figured it'd go through the window adjacent to the TV with no problem. Wrong.

The signal, it turns out, can't penetrate certain types of double-pane glass (such glass has some sort of gas between the panes that interferes with the signal). But a metal window screen was no problem. So, I left the window open when I wanted to monitor my horses. Not a problem in the summer, but when winter comes, the family squawks.

The only other option was to put the entire receiver outside, but it wasn't meant to be out in the elements, and putting it out there would eliminate the capability of the remote. Bill Thiel, owner of Saddlebrook BarnCams, tells me he's designed a new receiver, due out any day. It'll permit the antenna to be separated from the receiver and installed outside, and will operate by remote control. As soon as it's available, I'm told I can trade up. This company is very accommodating.

The other minor complaint was about sound quality. The color cameras transmitted excellent sound and allowed me to hear my horses chewing, walking around, sighing, and passing gas (music to any horseman's ear). But the black-and-white camera's sound quality was badly muffled, obscuring the more subtle sounds of equine night life and making audio monitoring more like eavesdropping through a closed door. To my surprise, the sound quality from a very cheap, black-and-white Radio Shack camera was as good as Saddlebrook's color camera. That is, excellent. Why, then, did I take the ultra-cheap Radio Shack camera back? Because despite its superior sound quality, its picture was lousy, its lens produced a narrow view that gave my horse a lot of places to hide completely out of view, and in low light I could barely make out the outline of my horse in a sea of gray snow. The only way I could monitor the horse in that particular stall was to turn the lights up all the way, which clearly disturbed his sleep. I hadn't truly appreciated the quality of the Saddlebrook cameras until I saw just how inferior the image could be with a lesser camera.

Thiel tells me that they've replaced their black-and-white cameras with a new style that has both the excellent image quality and the excellent sound quality.

So that shoots down both of my complaints. I love this system.

The live webcam setup requires a little guidance if you're not a computer genius (which I'm not). In addition to your Saddlebrook BarnCam camera system, you'll need a video capture card for your computer (Saddlebrook sells these), webcam software (available on Saddlebrook's website), and an Internet connection. You'll also need to subscribe to a website that will host your camera images so you can view them when you're away from home. Saddlebrook recommends their new partner in this endeavor, CI HOST (www.cihost.com). I haven't tried this outfit, so I can't comment personally, but Thiel tells me that for a competitive price you can get more bandwidth than with many other live webcam hosts, which is important if more than one person will be monitoring for you. So be sure to compare traffic capabilities as well as cost.

Cost and contact information:
My wireless BarnCam metal barn system for black-and-white surveillance cost $399; the color system cost $449. Contact Saddlebrook BarnCam™ by calling (920) 474-7776, or go online at www.BarnCams.com.

Why The Perfect Stall Loves the Saddlebrook BarnCam™

- Excellent video and audio quality.
- Minimal installation thanks to wireless transmission that's not confounded by a metal roof.
- Remote control switching allows monitoring of several cameras without getting up.
- Wide angle lens; no blind spots.
- Live webcam capability.

In case you didn't know...

Biting insects such as flies are more than just pests that make you and your horse miserable; they can spread disease, cause skin lesions, introduce parasites, interfere with wound healing, and irritate eyes.

Plus, their annoying presence rationalizes the use of insecticides, repellents, and even feed-through larvacides, all of which have a negative impact on the environment.

Chapter 8
Odds & Ends

Customizing Perfection
Your experience with The Perfect Stall components likely will depend somewhat on where your facility is located.

If you're in the upper peninsula of Michigan, for example, the air in your stalls has a much higher relative humidity than does the air in stalls outside Las Vegas, Nevada, where the summers are hotter and infinitely drier. You'll need to run the electric aerator through your stalls daily to keep them fresh-smelling and ammonia-free, because you'll be relying more on *aeration* than on *evaporation*. In Vegas, the dry heat evaporates urine so well that you may not need to run the stall aerator through the Equidry Bedding any more than twice a week, maybe less.

On the other hand, the dry air and wind in Vegas might make it necessary to vacuum the stalls once a month because of environmental dust settling on the bedding, whereas in Michigan four times per year might be plenty.

The goal is to create the best possible maintenance program, at the least possible cost and with the least possible time spent cleaning stalls, for the best possible comfort for your horse, with a constant supply of fresh, clean air in the stall, free of allergenic dust and ammonia fumes. No matter where your horse facility is based.

I've tested The Perfect Stall products for two years, and I can say with confidence that minor adjustments to your daily maintenance routine, to match your regional climate (and whatever little quirks might characterize your facility), are all you'll need in order for The Perfect Stall to be the most economical, efficient, healthy, and environmentally-friendly way to stall your horses. If you're as picky about these benefits as I am, you'll never want to go back to ordinary bedding and "stall furniture." Even if you're not particularly picky about health and environmental issues, the money savings alone should convince you to go Perfect.

With that said, there are two situations in which you might need a little help in making a smooth switch to The Perfect Stall.

The OSS (old stinky stable)

One situation is the old stable that already stinks to high heaven because of years of accumulated urine in the subflooring. If you were to kick all your horses out to pasture and leave them out for a month, the barn might still smell of ammonia because of a well-established ammonia factory cranking out fumes.

For such a situation, there is a product called SUPPRESS, manufactured by the Westbridge company, that is touted to be *organic*, *toxin-free*, and capable of reducing ammonia fumes in poultry operations and feeder pig barns to as low as 40 percent. This could speed up the process of clearing the old smells out of the stall before you install your Equidry Bedding, and it'll make you happier with its performance. Pig and chicken facilities generate the worst ammonia fumes of any operations I've ever visited, so this is a product worth looking into.

You'll apply SUPPRESS liquid to the cracks and corners of your stalls where urine may have soaked in over the years. It comes in 1-gallon and 50-gallon quantities, and smells faintly of vinegar. Although I'd rather eliminate ammonia fumes naturally, by aeration and ventilation, than by using chemical neutralizers, I've found that when I mix SUPPRESS solution with an experimentally created ammonia source, it really does work. Put this option up your sleeve in case you need it someday.

Fly hoards

Your old established horse facility may also be overrun with flies. It might seem hopeless to cut the fly population down without maintaining some sort of chemical warfare, but I hope you'll trust my advice enough to first give non-insecticidal means a fair shot. Remember, any time you apply insecticides to kill insect pests, you're also killing the "good" bugs that prey on the "bad" ones— and it takes the "good" ones significantly longer to repopulate the facility. So, in the long run, you're giving the pests the upper hand. To be frank, the use of insecticides may be one of the reasons why the flies at your facility are so bad in the first place.

The Perfect Stall setup naturally cuts way down on fly habitat and attraction, and with a little help from some friends called Spalding Fly Predators, you'll see real improvement. Not overnight, mind you, but the pest fly problem definitely will be improved in a month, and even better in two months, and even better in a year. It'll take a little patience, but the payoff can be amazing. At my farm, which used to be a dairy farm and is situated next door to a range beef cattle operation, I can attest that my fly problem was extreme.

Six years ago I bit the bullet and took all my half-used bottles of insecticide and toxic repellent to the haz-mat guys at our local waste-transfer station. That same day, I started using

Spalding's little bugs, which don't bother mammals but have a voracious appetite for fly larvae in the manure pile.

The fly population did decrease significantly, especially when you consider that the winters here in the past 15 years rarely have gotten cold enough to kill off insect eggs and larvae waiting for spring. But despite the improvement, I still needed some sort of protection for my horses, so I invested in fly masks, fly sheets, and "natural" fly repellents, such as those that contain essential oils like citronella). These milder repellents were effective, even though we used to be the neighborhood hotspot for fly pests, because I'd made significant progress in cutting down the fly population.

When I switched to Perfect Stall components, and specifically Equidry Bedding, the fly population plummeted even further, within one summer. I continue to using Spalding Fly Predators. The fly population has dropped so low that nowadays it's no different here at the farm than it is at a randomly selected house on main street in downtown Coeur d'Alene. The neighbor's range cattle are still grazing and pooping on the land adjacent to ours, and they still break through my fence on occasion and leave deposits in the horse pasture, but there just aren't enough flies here to partake of the booty.

Cost and contact information:
A gallon of SUPPRESS costs about $25; bulk amounts are also available. Call the Westbridge Company at (800) 876-

2767, or go online at www.westbridge.com.

Spalding Fly Predators cost about $2.50 per 1,000 and are available in multiples of 5,000. At the recommended 1,000 per horse per month, for my 12 horses, for a 6-month fly season, my yearly cost for predators is about $200. Call Spalding Laboratories at (800) 920-5772, or go online at www.spalding-labs.com.

The Perfect Stall

Bibliography

Stabling is associated with airway inflammation in young Arabian horses. *Equine Vet J* 33[3]:244-9 2001 May. Holcombe SJ, Jackson C, Gerber V, Jeffcoat A, Berney C, Eberhardt S, Robinson NE.

Sawmill Receives $20,000 Fine For Leaching Waste. *Fisheries and Oceans Canada* NR-PR-01-046E. May 10, 2001.

Airway inflammation and mucus in two age groups of asymptomatic well-performing sport horses. *Equine Vet J.* 2003 Jul;35(5):491-5. Gerber V, Robinson NE, Luethi S, Marti E, Wampfler B, Straub R.

Woodwaste Bedding. *Environmental Guidelines for Horse Owners:* Ministry of Agriculture, Food & Fisheries; Resource Management.

Woodwaste Use -- Precautions To Horse Owners. *Environmental FACTSHEET;* Ministry of Agriculture and Food; British Columbia 655.000-2. January 1999.

Coliform Counts in a New, Long-term Bedding Material Over Time. Unpublished; 2003. Hayes Karen E. N. DVM, MS. www.hayesk@earthlink.net.

Restrictive Proposal On Water Act Rules. *Stable Management* Oct 2001; 10-12.

Effect of Sodium Bisulfate on Ammonia Levels, Fly Population, and Manure pH in a Horse Barn. *Proceedings of the 42nd annual Convention of the American Association of Equine Practitioners* 1996; 306-307.

Moving Beyond Straw. McFarland C. *Thoroughbred Times* 19:25 June 21, 2003; 17-21.

Composting of Horse Manure. Alberta Agriculture, Food and Rural Development 99. Jerry Leonard, University of Alberta June 1999.

A Guide To Composting Horse Manure. Paige J; WSU Cooperative Extension, Whatcom County.

CAFO Final Rule. National Pollutant Discharge Elimination System (NPDES); U.S. Environmental Protection Agency. December 16, 2003.

Calorimetric determination of inactivation parameters of microorganisms. *J Appl Microbiol* 2002;93(1):178-189. Lee J, Kaletunc G.

Effect of Phone Book Paper Versus Sawdust and Straw Bedding on the Presence of Airborne Gram-Negative Bacteria, Fungi and Endotoxin in Horse Stalls. *J Equine Vet Sci* 18(7): 457-461. Jul 98 Tanner MK, Swinker AM et al.

Composting Characteristics of Three Bedding Materials. *J Equine Vet Sci* 18(7):462-466 Jul 98. Swinker AM, Tanner MK et al.

Quality of Different Bedding Materials and Their Influence on the Compostability of Horse Manure. *J Equine Vet Sci* 21(3):125-130 Mar 01. Airaksinen S, Heinonen-Tanski H et al.

Horse Hair Coat Cleanliness is Affected By Bedding Material. *J Equine Vet Sci* 17(3):156-160 Mar 97 McClain J, Wohlt JE, et al.

Alternate Bedding Materials for Horses. *Equine Pract* 17(1): 20-23 Jan 95. Thompson Kent.

Points of the Horse, 7th ed. New York: Arco Publishing Co., 1968; pp 223, 227, 515.

Inhaled endotoxin and organic dust particulates have synergistic proinflammatory effects in equine heaves (organic dust-induced asthma). *Clin Exp Allergy*. 2003 May;33(5): 676-83. Pirie RS, Collie DD, Dixon PM, McGorum BC.

A case control study of factors and infections associated with clinically apparent respiratory disease in UK Thoroughbred racehorses. *Prev Vet Med*. 2003 Jul 30;60(1):107-32. Newton JR, Wood JL, Chanter N.

Equine recurrent airway obstruction: pathogenesis, diagnosis, and patient management. *Vet Clin North Am Equine Pract.* 2002 Dec;18(3):453-67, vi. Davis E, Rush BR.

Inhaled endotoxin and organic dust particulates have synergistic proinflammatory effects in equine heaves (organic dust-induced asthma). *Clin Exp Allergy.* 2003 May;33(5) 676-83.Pirie RS, Collie DD, Dixon PM, McGorum BC.

About the Author

Karen E. N. Hayes is the award-winning author of hundreds of articles in such magazines as *Equus* and *Horse & Rider*. This is her fifth horse-care book (see page 144). A 1979 graduate of the University of Illinois College of Veterinary Medicine, Karen was in private practice for several years, joined the faculty at Wisconsin's veterinary school, earned a post-doctoral master's degree in equine reproduction there, then returned to private broodmare practice. She and her husband live in northern Idaho on a 100-acre farm (Ironhorse Friesians; www.ironhorsefriesians.com). They've raised Friesian horses since 1986. Karen invites you to keep up with the latest stall improvements and innovations by checking www.theperfectstall.com.

OTHER BOOKS BY KAREN E. N. HAYES

• **The Complete Book of Foaling**.
A comprehensive, step-by-step, hands-on guide for foaling attendants. ISBN 0-87605-951-5

• **Emergency!**
The Active Horseman's Book of Emergency Care.
Advanced first aid for the experienced performance horseman in a crisis, when immediate veterinary care is not available. ISBN 0-939481-42-1

• **Hands-On Horse Care**.
Basic, detailed horse care, from daily observations to advanced procedures such as bandaging, medicating eyes, and giving shots, and a unique step-by-step method for identifying the cause of the signposts a horse exhibits when injured, ill, lame, or just being a normal horse. In cooperation with the American Association of Equine Practitioners and winner of the 1997 American Horse Publications Book Award. ISBN 0-86573-861-0

• **Hands-On Senior Horse Care**.
How to keep your aging horse younger and more active, for a longer, more fulfilling life. ISBN 192916411-4